Origins

·Reconsidered·

IN SEARCH

OF

WHAT MAKES

US

HUMAN

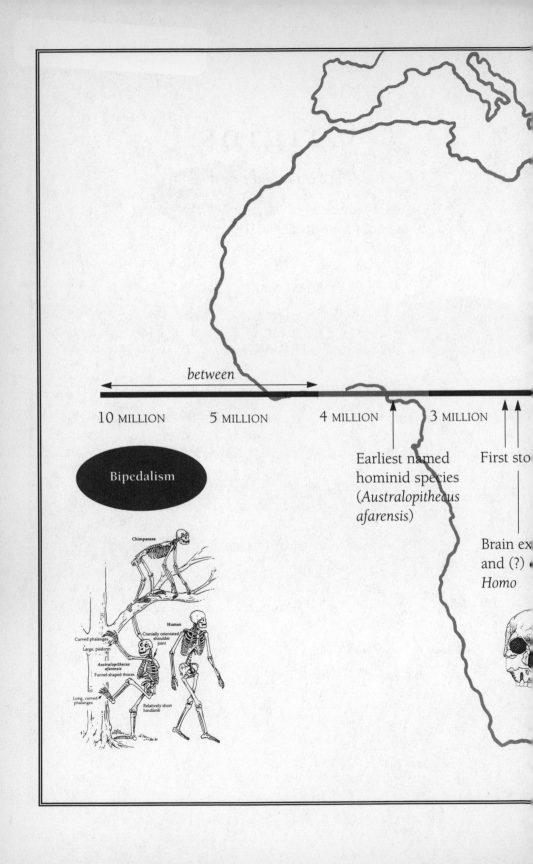

between

10 MILLION   5 MILLION   4 MILLION   3 MILLION

Bipedalism

Earliest named
hominid species
(*Australopithecus
afarensis*)

First sto

Brain ex
and (?)
Homo

Chimpanzee

Human

Cranially orientated
shoulder
joint

*Australopithecus
afarensis*
Funnel-shaped thorax

Curved phalanges

Large, pisiform

Long, curved
phalanges

Relatively short
hindlimb

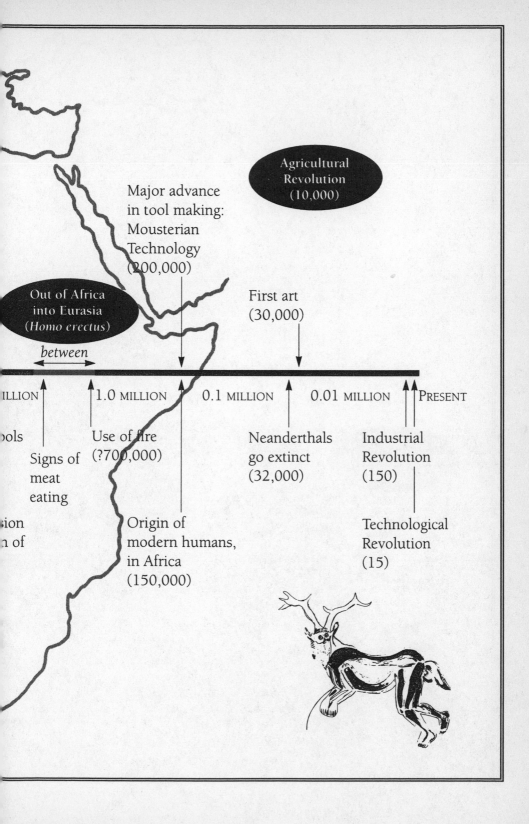

Out of Africa
into Eurasia
(*Homo erectus*)

Major advance
in tool making:
Mousterian
Technology
(200,000)

Agricultural
Revolution
(10,000)

First art
(30,000)

between

MILLION   1.0 MILLION   0.1 MILLION   0.01 MILLION   PRESENT

ools

Signs of
meat
eating

Use of fire
(?700,000)

Neanderthals
go extinct
(32,000)

Industrial
Revolution
(150)

sion
n of

Origin of
modern humans,
in Africa
(150,000)

Technological
Revolution
(15)

Origins

·Reconsidered·

IN SEARCH

OF

WHAT MAKES

US

HUMAN

Richard Leakey

and

Roger Lewin

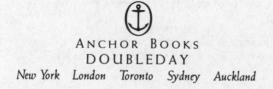

ANCHOR BOOKS
DOUBLEDAY
New York   London   Toronto   Sydney   Auckland

An Anchor Book

PUBLISHED BY DOUBLEDAY

a division of Bantam Doubleday Dell Publishing Group, Inc.
1540 Broadway, New York, New York 10036

ANCHOR BOOKS, DOUBLEDAY, and the portrayal of an anchor
are trademarks of Doubleday, a division of Bantam Doubleday
Dell Publishing Group, Inc.

Origins Reconsidered was originally published in hardcover by
Doubleday in 1992. The Anchor Books edition is
published by arrangement with Doubleday.

Book design by Claire Vaccaro

Library of Congress Cataloging-in-Publication Data

Leakey, Richard E.
Origins reconsidered :
in search of what makes us human / Richard Leakey
and Roger Lewin.—1st Anchor Books ed.
p.   cm.
Includes index.
1. Anthropology.   2. Human evolution.   3. Man, Prehistoric—Africa, East.
4. Africa, East—Antiquities.   I. Lewin, Roger   II. Title.
GN31.2.L4   1992
573.2—dc20   92-6661
CIP

ISBN 0-385-46792-3
Copyright © 1992 by Sherma B.V.

10   9   8   7   6   5

For Meave and Gail

Acknowledgments

Many years of work and frequent interaction with many colleagues are effectively distilled in the pages of this book. To recognize individuals among them would be invidious, because of those not named. We thank them all, and they know who they are. Two cannot remain anonymous, however; they are Kamoya Kimeu and Alan Walker, long-time friends and inspiring colleagues.

The government of Kenya and the Governors of the National Museums deserve special mention for enabling and encouraging our search to proceed.

Finally, we thank our spouses for their strong and steadfast support.

Contents

xi

Contents

Prologue

For more than two years now, I have lived with constant reminders of danger: armed soldiers guard my house, bodyguards accompany me in my Land Cruiser, and others follow in another vehicle, wherever I drive. I am surprised by how quickly I have become used to their presence, a part of everyday procedure. But I never quite forget that there are people who would rather see me dead than alive.

In April 1989 President Daniel arap Moi, Kenya's head of state, surprised me and a great many others by appointing me director of the Kenya Wildlife Service. My job was to put an end to the rampant poaching of elephant and rhinoceros and to establish a structure for managing wildlife, the focus of our tourist industry. This industry is vitally important to Kenya because of the foreign currency it brings into the economy, but by preventing ivory poaching I was thwarting powerful people who had been making a lot of money from the awful slaughter. That was why they wanted me out of the way.

Now I am immersed in an ambitious program to encourage wildlife and human populations to thrive side by side. The balance will be difficult, given the press of growing human populations and the fragility of diminishing communities of wild creatures. In many ways it is a microcosm of a predicament faced worldwide. I plan to succeed.

When President Moi called on me to take over the job, I considered it an honor to accept. I was aware of what I was taking on and also what I was leaving behind. For twenty years I had been director of the National Museums of Kenya and had spent as much time as I could visiting Lake Turkana, in northern Kenya, looking for early human fossils. Fossil hunting was and remains my first love.

I am fortunate to live and work in the continent that Charles Darwin called "the cradle of mankind." I am fortunate, too, to have grown up with a family tradition of independence, determination, and the knowledge that even the most hostile physical environment need not be life threatening. Wild nature was as familiar to me as kindergarten and high school were to most youths. I can survive where many Westerners would succumb to thirst and hunger or to predators. I learned these skills as a child.

One doesn't have to be an adventurer in the savannah to go into remote areas seeking the fossil remains of our ancestors. But knowing how to find food, where to look for water, and how to avoid danger in an arid and bare landscape has given me a sense of peace, a sense of belonging. I feel a connectedness with our ancestors, an intimacy with a land that was their land. And, of course, there is the Leakey tradition. My parents, Louis and Mary, revolutionized the search for human origins with their world-famous discoveries.

Although as a youth I fiercely wanted my independence and desperately fought to be out of my parents' shadow, I was drawn into the search for our beginnings, the understanding of what made us what we are. Even now, I find it difficult to explain how the emotion underlying that quest overcame my intellectual de-cision—frequently expressed—not to have anything to do with fossil hunting. Save, of course, for the adventure, the challenge of being out there in the wild. Louis died suddenly, in 1972, and I'm glad to say we had reconciled our differences. He had come to terms with my independence, and I with the fact that he had made major contributions, which I had previously failed to see or understand.

Long after I became involved in fossil hunting, but while my father and I were still clashing antlers, I came across a manuscript of a lecture he had given, in California, I think. One sentence arrested my attention: "The past is the key to our future." I felt as if I were reading something I had written; it expressed my own conviction completely. Had we come to this realization sepa-rately? Or had I unconsciously absorbed it from him? I doubt the latter, because as a boy I was not much interested in what he was doing. My father was religious, although not in the conventional sense; I am not. Yet apparently we had come to the same numi-nous point. It was a poignant moment for me, seeing those words written in my father's hand: "The past is the key to our future."

During the years of organizing fossil hunting at Lake Turkana I was aware of more than the experience of discovery; I found within myself a certainty about what I was about to learn. I felt that there, in the arid sediments around that magnificent lake, answers were to be pieced together that went beyond the ques-tions normally asked in science. If we could understand our past, understand what shaped us, then we might gain a glimpse of our future. Humans worldwide belong to one species, *Homo sapiens,* the product of a particular evolutionary history. It is my convic-tion that an understanding of that history can inform our future actions as a species. Most particularly, we need an appreciation of our relationship with the rest of the natural world.

There is a deep personal drive behind the search for human origins. True, paleoanthropology can be as technical in its approach as many another scientific discipline: from statistical analysis to the arcane data of molecular biology, the pursuit of human origins is intellectually demanding and rigorous. But it is more than that. Because the target of the search is ourselves, the enterprise takes on a dimension absent from other sciences. It is in a sense extrascientific, more philosophical and metaphysical, and it addresses questions that arise from our need to understand the nature of humanity and our place in the world.

Each time I give a public lecture, I am reminded of this need to know about ourselves. The audiences that come to hear me are, I often feel, seeking a kind of reassurance. I talk about fossils and anthropological theories, and the people ask me whether monkeys can know the meaning of sin and what will happen next. Once, about ten years ago, an elderly lady, clearly concerned, wanted me to tell her whether it was true, as she had heard, that "humans are only a historical accident." I could tell her about Earth history and the fossil record. I could discuss chance and evolution. And I could describe alternative worlds, without humans, perfectly plausible worlds. But it was clear that she really wanted to be told that no, humans are not a biological accident; *Homo sapiens* was bound to happen. Her humanness, her urge to make sense of her world, seemed to demand that it be so.

Paleoanthropology, therefore, has a mixture of scientific and extrascientific elements. Most of the time, of course, we professionals are concerned with the bones themselves: how anatomical constellations in one cranium may relate to similar constellations in another, the two perhaps being separated by a million years of evolutionary history. It is an absorbing occupation, one that tests our abilities to recognize genetic links in the most meager of physical evidence. The philosophical element is always present, but usually as an unspoken rhythm to our work.

Fifteen years ago I decided to write a book, with Roger

Lewin, about where I saw the state of paleoanthropology, and also to address some of the philosophical issues that concerned me then. Just recently, in a rare quiet moment, I sat down to read *Origins* again. I see that its main philosophical message was that, contrary to much popular wisdom, the human species is not unswervingly aggressive, not genetically driven to violence. Many prominent figures, including Konrad Lorenz, had argued that territoriality and ritual combat in animals, extrapolated to the human arena, explained the penchant for warfare that has so marked recent human history. Other prominent figures, most notably Raymond Dart, suggested that evidence of bloody combat could be discerned in the human fossil record. Together, these two lines of evidence were welcomed with a curious enthusiasm by a public that seemed eager not only to explain warfare, but also to explain it away.

As we showed in *Origins*, both lines of evidence were flawed. Territoriality is a flexible behavioral trait in many animals, often influenced by ecological circumstance. And human behavior, of course, is flexible in the extreme. Humans do not march in lockstep to the demands of aggressive genes. Our behavior as a species is complex, always shaped by cultural context and always amenable to choice, to free will. We argued that a willingness to accept the notion that *Homo sapiens* is driven to violent conflict by biological imperative is itself a cultural manifestation. To suggest that warfare is common in human history because of our genetic heritage absolves us from guilt: one cannot fight the inevitable, or so the argument went. But our position in *Origins*—that conflict is in the realm of free choice—places responsibility on human society, a burden some apparently would prefer not to bear. I believe that we do have to bear that responsibility, both for what happened in the past and for what the future may hold.

When, in *Origins*, we examined the putative evidence for violence in prehistory, the second line of argument popular at the time, it wilted under the sharp gaze of scientific investigation. Skulls said to have been crushed by lethal blows in fact had been damaged by natural processes during fossilization. What were thought to be weapons turned out to be nothing more than the remains of hyenas' dinners. There is no evidence of frequent violence or warfare in human prehistory until after about ten thousand years ago, when humans began to practice food production—the leading edge of agriculture.

We argued that evolutionary history has endowed our species with an inclination to cooperate. In addition, *Homo sapiens* has a greater flexibility of behavior, a broader range of choice of action—and therefore of responsibility—than any other species. Much of the conflict in the world can be traced to materialism and cultural misunderstanding, not to our biological nature. With one's own possessions to protect or another's to covet, one can find material advantage in military conflict; no doubt about it. History has demonstrated that repeatedly. But we are not driven by some kind of inner demon to fight one another, as Lorenz and Dart believed. We have a choice in the matter, and a responsibility.

More in the scientific realm, I see a range of anthropological ideas and interpretations presented in *Origins* that have turned out to be wrong. As philosophers constantly tell us, and we keep finding out the hard way, science is tentative; existing perceptions are constantly replaced by new ones. And so it will be in the future, because this is the way of scientific progress. But I hope, too, that fifteen years of experience have helped me to be less inclined than I once was vigorously to defend conclusions, to insist that what we think we know now is the Truth. Absolute truth is like a mirage: it tends to disappear when you approach it. One of the most important lessons for me during these years is my learning that, passionately though I may seek certain answers, some will remain, like the mirage, forever beyond my reach.

. . .

Instrumental in overturning some of the ideas and interpretations we wrote about in *Origins*, of course, are new fossil discoveries, some my own, some the work of others. The last fifteen years have been extraordinarily productive in terms of fossil finds, often in ways we did not predict. In 1968 I had started explorations among the vast sandstone deposits on the eastern shores of Lake Turkana. I was fortunate enough to make discoveries that propelled me to the kind of fame my father had enjoyed, a matter of some satisfaction, I have to admit. Often I would gaze across the jade-green waters of Lake Turkana, wondering what secrets the sediments of the western shore held.

But my plans to explore that shore were interrupted by several happenings, among them the complete failure of my kidneys in 1979. A normally simple bacterial infection a decade earlier had insinuated itself into my kidneys and begun a slow process of deterioration. My doctor told me that one day my kidneys would fail, and I should expect to die young. I decided the only way to proceed was to put it out of my mind, ignore the prospect, and tell no one. In the end, however, the inexorable pathology cornered me, and I railed against its effects, against anything that could stop me from doing what I so much wanted to do. In July of 1979, in the final stages of kidney failure—the characteristic deep, steely cold of my body, the overwhelming nausea, the mind-draining fatigue—I flew with Meave, my wife, to London for treatment. I left Kenya in silent tears, wondering whether I would ever see my home again, my friends, my family. I wondered whether I would see Lake Turkana again, make the discoveries I knew were there to be made.

I was lucky. A transplant from my younger brother, Philip, effectively gave me a second life. Feeling that whatever years I now had ahead of me were a bonus, I initiated the exploration of the western shore of Lake Turkana. The discoveries were worth waiting for, as I shall recount. Following our first tentative prospecting came remarkable finds, some of them technically stunning, some emotionally thrilling.

During these years other shifts were taking place in paleoanthropology, several based on evidence from archeology, several on molecular biology. Together, they touched on a topic that has become prominent in recent years: the origin of modern humans, people like us. The evolutionary history of humans necessarily encompasses the reconstruction of, for want of a better phrase, the family tree. From this we begin to see the origin, transformation, and extinction of a group of human-related species through time, eventually leaving extant just one species, *Homo sapiens.* All stages of the history of the tree have their fascinations, but the last stage—the origin of modern humans—is salient. Here, we are witnessing the final emergence of people of our own kind, people imbued with the quality of humanness that we experience today.

For me, the emphasis on the origin of modern humans is stirring because it relates to the philosophical issues that run deep through our profession. Much of the evidence is in fragments of anatomy, enigmatic archeological assemblages, and DNA gels in molecular biology laboratories. But when we apply this evidence to the questions of how, where, and when modern humans arose, we raise more metaphysical questions. What is the origin of our humanness? What do we mean by humanity? Addressing the scientific issue of the origin of modern humans forces us to think about our existence as individuals and as a special species.

For centuries philosophers have dealt with aspects of humanness, of humanity. But, surprisingly, there is no agreed-upon definition of the quality of humanness. It hardly seemed necessary, partly because it appeared so obvious: humanness is what we *feel* about ourselves. Those who tried to define humanness found themselves molding Jell-O: it kept slipping through the fingers. But if this sense of humanity came into being in the course of evolutionary history, then it must have component parts, and they in turn must be identifiable. It is my conviction that we are beginning to identify these components, that we can see the gradual emergence of humanness in our evolutionary history. I am therefore perplexed by, and impatient with, a popular alternative view that is championed by several scholars.

These people suggest that the quality we call humanness sprang fully formed into the brain of *Homo sapiens*. Humanness, according to this view, is something recent in our history, something denied to any of our forebears. By proposing that this special quality we experience as individuals appeared out of nowhere, so to speak, unconnected to our evolutionary heritage, these people effectively make humanness a unique and scientifically inexplicable mark of humanity. This position casts a cloak of mystery over the very thing we are most urgently seeking, and in a curious way smacks of a kind of creationist obfuscation. I strongly reject it. I believe that the qualities of the human mind, like the form of the human body, have been shaped by a fascinating evolutionary history. It is the business of paleoanthropologists to reconstruct that history, not to obscure it.

With all these developments taking place, my own excavations, and the new focus on modern human origins, I concluded a little more than three years ago that it was time to write a new book, time to put the evidence into a new perspective. In particular, it was time to put into larger perspective this very special species *Homo sapiens* and what we can say about its place in the universe of things. Without claiming to have identified the Truth, I felt that a more penetrating glimpse in that direction was now possible.

Then came the call from President Moi. Since my energies and time were diverted to the exigencies of saving the elephant and other elements of Kenya's wildlife, I thought the book would have to wait; my ideas could not be sufficiently focused. What happened, though, was that, faced with the problems of wildlife conservation, I found a greater clarity suffusing the questions I had raised for myself. Samuel Johnson once said, "Depend upon it, sir, when a man knows he is to be hanged in a fortnight, it concentrates his mind wonderfully." Well, the presence of body-guards had a similar effect on my mind.

More pertinent, however, was the nature of my new preoccupation: understanding the interaction of human populations with wildlife communities. This was a second perspective on the place of our species. Rather than blurring my thoughts, it sharpened them. Rather than diverting my attention, it provided me with a lens through which to focus it. I decided that I would go ahead, conscious that I am doubly privileged to have been involved in the search for the fossils of our ancestors, and to be engaged in the fight to reconcile the press of human populations on communities of wildlife.

Origins Reconsidered is a personal journey of discovery. Working with Roger Lewin, I hope to share some of this experience, beginning with the first ventures to the west side of Lake Turkana, the first extraordinary excavations there. The experience comes in many parts, the mundane and the sublime. The day-to-day practicalities of finding the right spot for the camp-site, of locating a water source in an otherwise dry terrain, the frustrations of having promising fossil finds turn out to be barren —these are the beginnings of the quest, the unglamorous but unavoidable first steps in the long journey. Then there are the discoveries themselves, the cause of camp celebration, the stuff of newspaper headlines—ultimately, of course, the fragmentary remains of individuals with whom we have a genetic link, no matter how tenuous, the parts of an evolutionary puzzle we are slowly piecing together. The west side discoveries were all of these and more.

PART ONE

In

Search

of

the

Turkana

Boy

To West Turkana

T h e y   h a d set out early, this band of six purposeful individuals, striding across rolling, grassy terrain punctuated here
and there by flat-topped acacia trees. The sky hovered between
gray and pink as the sun rose close to breaking the line of hills in
the east, on the other side of the vast lake. Soon the mountains
in the west would be fringed with the colors of morning. The
soft air carried the scent of the great waters on a predawn
breeze. Herds of three-toed horses and gigantic wildebeest were
already drinking along the sandy beaches, long drafts of silky
water that would charge their needs for the day. Wading birds
scurried delicately among the waves at the lake edge, expertly
plucking tiny fish and shrimp from the jade-green water. Above,
thousands of flamingos wheeled in pink flocks, swooping
through the air in an exuberant greeting of another African day.

Everyone had heard the saber-toothed cats during the night,
repeated choruses of throaty moans, a sure sign of a hunt in
progress. Even though the band felt itself relatively safe at its

riverside camp a mile from the lake, there was always tension when saber-toothed cats were near. Only a year ago a child had been attacked when he strayed from the watchful eyes of his mother and her companions. Returning hunters, the same group of men who were setting out this day, arrived just in time to drive the predator away. But the boy had died some days later from the loss of blood and the kind of rampant infection that can be so deadly in the tropics. Not surprisingly, this morning's discussions urged extra care on the women and their offspring, gathering tubers and nuts near the camp, and the men on their hunt. These men too were predators.

A herd of large antelope, with sleek brown hides and corkscrew horns, was the day's target. Signs of the herd had been seen yesterday, and if calculations were correct, today it should be about fifteen miles to the north, a short stretch for these hunters, their powerful, athletic bodies fit to cover long distances with ease. Brothers and cousins all, including the youngest boy. Despite his youth, he too was tall, slender, and muscular, his face broad, marked by a low, sloping forehead and prominent brow ridges, like those of his kin. It was to be his first hunting foray. And his last.

Expert trackers, the hunters concentrated on detecting signs of their quarry. Unlike the garrulous females back at camp, they had little need for lengthy exchanges. They were subdued, succinct. Hunting demanded quiet, an ability to merge into the terrain. Camp life and food foraging encouraged discourse; the camp was a secure place, a place for free communication, a place for extensive learning for the young. It was a noisy place, with intense socializing among young and old, where games were played for pure fun and for social advantage.

By noon the herd was in view, quietly resting near some shade trees, the animals' strategy to stay cool. There would be no other predators active at this time of day. On their way the hunters had spotted a group of large primates, bipeds like themselves, but bulkier and with massive jaws. These bipedal primates weren't hunters; they foraged for plant foods, including tough fruits on the trees and bushes of the woodland and patches of forest. They scurried away when they saw the hunting band approach. They weren't hunters, but occasionally they were hunted, so they were properly nervous.

Later the hunting band saw a tight herd of a kind of elephant, with enormous, elaborate tusks. The hunters would have gladly scavenged a newly dead carcass, but none was apparent. As it was, these animals were too big, too risky to tackle as prey. The antelope would be a surer, safer target. A young one would be chosen as today's prey, or perhaps an older, disabled animal. What these hunters lacked in natural weapons, they would make up in stealth and cunning. To an arsenal of crude short spears and rocks, the hunters added simple but effective traps and a knowledge of how to steer prey toward them. When the herd came into view through an acacia grove that screened the hunting band, a strategy was formulated. Someone indicated a cache of rocks that had been made nearby, rocks suitable for making the stone knives and axes necessary for butchering a carcass.

A target individual, a young one, was selected as prey, and the hunting band split up. Each member knew what he must do in the effort to split the herd and drive the prey animal toward the trap, a device made of hide, strips of bark, and branches. Perhaps it was because the herd was bigger than the hunters had thought; perhaps it was because the antelope, like the humans, were especially alert today because of the invisible presence of saber-toothed cats in the area; perhaps it was because the boy had a lot to learn and was nervous lest he fail in his role. Perhaps it was a combination of all these things that caused the plans to go awry. Whatever it was that happened, the boy suddenly found himself running, running blindly, a long gash on his thigh bleeding profusely but, curiously, with no pain. Not yet, anyway.

Weak from loss of blood, the boy became frightened as darkness fell. His wound hurt now, and throbbed. He remembered how, a year earlier, the child who had been attacked by the saber-toothed cat had wounds like this, inflicted by the predator's teeth, not, as had happened now to him, by the glancing slash of an antelope's horn. He remembered how the child became feeble, acted strangely, flailing his arms and shouting wildly. He remembered how, finally, the child stopped moving, was very still. He never saw him again. The memory was frightening, but he wasn't sure exactly why.

One day passed, then another. Where was everybody? Why didn't they come? If only he could get to the lake, he would feel better in its cooling waters. His whole body raged with fever. If only he could get to the lake. It wasn't far. Surely he could make it, and then they would find him.

The boy did make it to the lake's edge, a shallow lagoon with lush grass all around, reeds growing from the bottom. Crawling, he hauled his ravaged body into the soothing water, the fever close to claiming its victim. For a short while he did feel better, calmer, sleepier, very sleepy.

They never found him, in that shallow lagoon on the western side of Lake Turkana, a little more than 1.5 million years ago.

"Kilo November Mike, requesting take-off clearance." We were loaded up, fueled, all checks completed, and the single-engine Cessna was ready to go. I waited for the response, eager to be on our way. The OK crackled over the speaker, just above my head. *"Kilo November Mike, cleared for immediate take-off."*

I opened the throttle and felt the vibrations build in the fuselage as the power of the engine strained against the braked wheels on the stationary aircraft. One more visual inspection of the busy skies around Nairobi's Wilson Airport, and then I released the brake. 5Y–KNM surged forward as if it too were impatient to be on its way, and accelerated down the runway to the east, soon to leap into the early morning sky. A wide bank to the right pointed the nose in the direction of Lake Turkana, our destination, some 260 miles due north.

The flight to the lake takes just over two and a half hours. But it was really a journey back in time. Waiting for us at the end of the trip were a few fragments of an individual who had lived a little more than 1.5 million years ago: the Turkana boy. His kind were our ancestors, and the boy himself had a big surprise in store for us.

I don't care to remember how many times I've made this flight, but the landmarks are as familiar to me as any commuter's route to work. I first piloted myself here in 1970, in the early days of exploring the ancient deposits on the shores of Lake Turkana, before the discoveries of human fossils had made the area famous. And, except when I fell ill and needed time for surgery and recovery in 1980, I've done the Nairobi-Turkana-Nairobi loop several times every month, sometimes alone, often with colleagues and visitors. But always with thoughts about the fossils we had recently unearthed and how we could search for new ones. Aloft in a plane is a good place for thinking.

On this particular flight, August 23, 1984, Alan Walker and I were on our way to join several teams of colleagues who were surveying the fossil territory on the west side of Lake Turkana. Alan and I have been close friends and colleagues since 1969, when I invited him to describe the hominid fossils discovered on the first major expedition to Lake Turkana. A tall Englishman with an athletic build—if not, these days, an athletic inclination —and a direct manner, Alan is a brilliant anatomist and a talented sculptor. He is also a recent recipient of one of the prestigious MacArthur genius fellowships.

The day before the flight, when Alan was working on fossil ape skeletons at the National Museum in Nairobi, Kamoya Kimeu called me on the radio phone to say that fragments of

7

hominid skull had been found at two different sites. "You might want to see them," joked Kamoya, knowing that I would very much want to see them. Kamoya is the leader of the specialist team of fossil hunters—the Hominid Gang—and he checks in every few days by radio phone when he and his team are out in the field. This is sound security practice, and it also helps to maintain a camp far distant from a regular supply source.

When I asked him to tell me more, Kamoya described the finds: several small fragments of skull. They didn't seem to portend anything special, but hominid fossils are rare finds. "Keep them safe for me, and we'll see you tomorrow." We discussed camp business, various supplies and equipment that needed to be taken up to the lake, and signed off. Alan and I made preparations to leave for Turkana the following day.

"We'll need to check out that tuff at Lothagam," Alan reminded me as we left the Nairobi skyline behind. Tuffs, or layers of volcanic ash, are a godsend to anthropologists, because it is usually possible to date them by geophysical analysis. But sometimes they are intransigent, and the one at Lothagam was giving us trouble. "We've got to get the age of that thing pinned down."

Lothagam Hill lies, like a brooding lion, west of Lake Turkana. Eerily beautiful, spectacularly splashed with yellows, reds, and purples, Lothagam has long been an enigma. The low outline of the hill, with a raised ridge at one end—the lion's head and mane—belies a complicated geology that makes it difficult to estimate the age of some of its rocks. And the age is important, because in 1966 a piece of hominid jaw was found there, and the date of the rocks would help us to infer the age of the fossils in them. Twenty years later, we still weren't sure whether the fossil was 5.5 million years old, closer to four million, or even less. If the jaw really was older than five million years, it would be the most ancient hominid known, close to the beginning of human prehistory. Yes, we had to get that one sorted out somehow. "We'll take a couple of passes, see if we can see where the tuff goes," I replied to Alan. But that was still more than two hours ahead of us.

"Kilo November Mike. Zone boundary outbound. Clearance to contact center, over." The journey north began with a mandatory check-in with flight control. *"Roger, Kilo November Mike. Over."*

We were ten minutes out of Wilson Airport, seven thousand feet and still climbing. As usual at this time of year, there was a good deal of cloud cover. We would lose it as we got farther north, but here it could make a tricky part of the flight just that much more difficult. We needed to gain altitude, quickly, because directly in front of us was the shoulder of the Great Rift Valley. We were thirty miles out of Nairobi.

The city of Nairobi lies at five thousand feet, resting on a vast geological dome that fifteen million years ago heaved the continental crust up from near sea level to more than nine thousand feet at its highest point. Under strain from deep mantle movements, the continental rock stretched and eventually gave way, forming the so-called Gregory Rift, a geological gash that stretches more than three thousand miles, from Israel in the north to Mozambique in the south. A geological episode of unimaginable proportions, the formation of the rift played a vital role in the evolution of our species. In fact, it is possible that had the Gregory Rift not formed when and where it did, the human species might not have evolved at all—ever.

But of immediate concern was that the shoulder of the rift ahead of us rises up to almost nine thousand feet. Any pilot going north out of Nairobi has to negotiate it, and several have failed. I was the pilot and needed to concentrate on the business at hand. I've noticed the occasional nervous passenger seeming to will the aircraft to a higher altitude. I'm grateful, of course, for all the help I can get.

Small and large farms, tea and coffee plantations, form a verdant patchwork on the fertile highlands to the east, around

9

the town of Limuru. The volcanic soil here is red and fecund. No wonder the British chose to settle here when they colonized the country a century earlier. Not far away is Thika, the home of Elspeth Huxley, who recorded her early life in Kenya in several books, including the famous *Flame Trees of Thika*. My attention, however, remained dead ahead, as the small aircraft was buffeted by the winds gusting up from the valley. We breasted the shoulder of the rift, and the sun briefly broke through the clouds here and there, but in a watery way. To the west, the walls of the Rift Valley drop away precipitously. Under clear skies the drama and contrast are plain to see, the green highlands giving way to the parched valley bottom. But this day, under an almost continuous cloud cover, the valley was in misty gloom. It is never the same, month by month, and I like that.

Often by this time in the flight it is possible to relax a little. Never completely, not least because of the danger of colliding with large birds, such as vultures, hawks, and even pelicans. A plane that hits one of those thirty-pound objects at 150 miles per hour can be in serious trouble: a hole in the fuselage or a broken propeller. A pilot who finds himself in the middle of a flock of the birds faces the almost impossible task of dodging them. I got very close once, and was lucky to avoid them. It may sound bizarre, but birds are the greatest hazard to the bush flier's life. I didn't see any potential collisions on this flight, but the low clouds made it a rough ride. So I tried to climb above them, eleven thousand feet now, the nose up—too far. The stall beeper —often a source of alarm to those not used to flying in small planes—sounded urgently, and I had to dip the nose to gain air speed. Stable again.

To the east, the peaks of the Aberdare Mountains stood above the clouds. Nourished by abundant moisture, the lush vegetation of the Aberdares supports a wonderful diversity of wildlife, including elegant black-and-white colobus monkeys and even some leopards. There used to be tens of thousands of elephants, too, but regrettably, no longer. About fifteen hundred of these majestic beasts live in the Aberdares Park now, protected from poaching. Beyond the Aberdares to the east is Mount Kenya, rising to a snow-covered peak at just over seventeen thousand feet. This day it was invisible to me, mantled in cloud. Although I couldn't see the mountain and its broad apron of fertile slopes, I thought again of the contrasts that surrounded me: mountaintop glaciers, alpine meadows, and thick temperate forest on Mount Kenya to my right, parched desert on the rift floor to my left, and a complex and graded mosaic of vegetation linking them. Even if one were not passionate about nature, one could not fail to be impressed by the vitality and the diversity of it all.

An hour out of Nairobi, still trying to find a smooth ride, sometimes above the clouds, sometimes below them, we could now see Lake Baringo over to the west. One of the many lakes scattered along the floor of the Great Rift Valley, Baringo is muddy brown at this time of the year, the result of seasonal heavy rains that bring silt down from the Tugen Hills to the west of the lake. The volcanic island in the center of the lake stands out clearly against the brown waters. My elder brother, Jonathan, lives close to the lake, on its western shore, where he grows melons and once raised snakes. Geologists and anthropologists have been prospecting for fossils for some years at various sites between the lake and the western hills, with considerable success. They are building up an exquisite picture of the animal life that abounded in the area some five to thirteen million years ago. Nothing spectacular in the hominid line yet, but you never know. For no very good reason, Baringo has never had a particular attraction for me: I much prefer the wilder terrain farther north.

"Kilo November Mike. Reporting operations normal. Over." That was the last check I'd make with air traffic control; soon we would be out of range. We were on our own now. In about forty-five

11

minutes we would sight the southern end of the lake. *"Roger, Kilo November Mike. Over."*

For much of the flight Alan and I usually don't talk much. You have to shout to make yourself heard above the engine noise, and that can quickly become uncomfortable. I don't mind the effort, but Alan, like most people, prefers not to shout. So he reads a book.

I've always enjoyed looking out of airplanes to see what is going on below: encroachment on forests, sedimentary outcrops, evidence of people, that sort of thing. The trip from Nairobi to Lake Turkana is particularly interesting because the terrain changes from south to north and from east to west like a geological kaleidoscope. Often, because the light is at a certain angle, you see something you've never seen before, a geological feature or a new water spring. I look at the ground a lot—when I'm not looking out for birds.

Until this point in the flight, the floor of the Rift Valley has been to the left as we fly above the Laikipia Plateau. But now our path takes us across the edge of the plateau, with the valley floor falling away beneath us, arid terrain from here to our destination. I love the desert. From here, for the last hour and a half of the flight, we see lavas and craters, dry lake beds, shadows of evanescent water courses in the dry earth. To some, such terrain seems hostile. To me, it is like coming home, and I feel a sense of peace. In the early morning the flight can be magical, with the sun's rays raking low across the landscape. Sometimes it is so beautiful that I feel I'd like to park the plane on a cloud and just watch it for a while.

I've always been passionate about wild and remote places and have had an intense interest and love of animals. As a teenager I wanted nothing more than to be a game warden, out in the wild, trapping dangerous animals, leading a life of adventure. And now that I'm director of the wildlife service, I'm boss of all the game wardens in Kenya. It is not hard to account for my love of the wild. My parents, Louis and Mary, did not allow the fact that they had infants or small children to interfere with their fossil-hunting expeditions: my brothers, Jonathan and Philip, and I were dragged along, often to the most exciting and dangerous places.

In the cool of the late afternoon and early evening, when excavation had finished for the day, Louis used to take long walks, searching for new sites or checking on old ones. He often took the three of us with him, with stern warnings that we weren't to dawdle. My father was a great naturalist, and he used to tell us about natural history as we walked. I was enthralled, and Jonathan, Philip, and I absorbed a deep knowledge of the countryside. We also learned how to fend for ourselves: how to find water and food in what looked like a barren desert, how to track and trap wild animals. We learned how to be part of nature, to respect it, not be afraid of it.

It wasn't all fun. Little boys can get bored when their parents spend hours scratching away in the dry earth. One day when I was six, fed up with what was happening—or, rather, as it seemed to me, not happening—I began complaining about being hot, thirsty, and generally uncomfortable. At last, Louis could take no more. He said, "Go and find your own bone!"

I wandered off, looking for likely prospects—though I'm not sure now what exactly I was looking for—and saw a piece of brown fossilized bone protruding from the ground, about thirty feet from where my parents were working. I set to, earnestly excavating the bone, and I became so engrossed that my parents got suspicious. When they saw what I had, they rapidly displaced me so that the fossil could be recovered without damage. It turned out to be the first complete jaw of an extinct species of giant pig, *Notochoerus andrewsi*, which had lived a half-million years ago. Yet no amount of praise for my discovery could expunge the fury I felt at having my bone taken away from me. I was fiercely independent even then, so much so that it wasn't long before I decided that whatever career I was to follow, I would certainly not be a fossil hunter. I would not follow in my parents' footsteps and be in their considerable shadow.

Louis and Mary will be forever associated with Olduvai Gorge, in Tanzania, site of famous discoveries that put East Africa on the anthropological map. Since 1925, South Africa had been the focus of the search for early human ancestors, and great successes were achieved by Raymond Dart and Robert Broom, legends in the annals of anthropology. But in East Africa at the time, nothing had been discovered. Then, after years of unrewarded searching, Louis and Mary made two major finds in quick succession, in 1959 and 1960. First was *Zinjanthropus*, a large-toothed, extinct species of hominid similar to some of the fossils unearthed in South Africa. Then there was *Homo habilis*, which Jonathan found. This was a new species of fossil human—a tool maker, big-brained, a member of our genus, and, according to my father, the direct ancestor of later humans. *Homo habilis* actually means "handy man," a name suggested by Raymond Dart.

The pattern of human prehistory that these early discoveries established is essentially with us today. From the earliest times, there were small-brained, bipedal apes, which included *Zinjanthropus* and the South African creatures, various species of *Australopithecus*. These eventually became extinct, and at some point there emerged the large-brained species, which became us, the genus *Homo*. These days we have a much clearer picture of the times in our prehistory when the various players in our ancestry put in their first appearance, and when—as in the case of most of them—they disappeared. At the time my parents were working at Olduvai Gorge, however, only a thin slice of this history was visible. Nevertheless, it was clear that *Zinjanthropus* and other small-brained creatures like it had lived close to two million years ago, and possibly earlier. By a million years ago they had become extinct. What was surprising was that *Homo habilis*, the first species in the line that leads to us, also originated early, at least as early as *Zinjanthropus*.

Homo habilis was exactly what Louis had been looking for, what he knew he must one day find: evidence that humanity—*Homo*—has deep roots in evolutionary history. The idea was something of a tradition in the British anthropological circles of the 1920s and 1930s, and Louis absorbed it from his mentors. His unique contribution was to put flesh and blood on the idea by finding a spectacular fossil. Without fossils, even the best ideas don't thrive. But fossil discoveries—more particularly, their interpretations—can be controversial. So it was with *Homo habilis*. And so it was to be with a similar fossil discovery I made, a little over a decade later. I was musing on some of this as we continued our flight north, and I wondered about Kamoya's new fossil discoveries.

Although we were now only two hours out of Nairobi, we could have been in another world. Everything was parched, the high plateau to the east, the desert below and to the west. This is my kind of country. We had reached the southern tip of Lake Turkana, which was way over on our right; its path south had been dammed by volcanic uplift some ten thousand years before. Fifteen miles into the lake from the southern shore is South Island, the site of many local legends. Fed by the massive River Omo, which drains the Ethiopian highlands, the lake has a fascinating geological history, only now being understood. In its more recent past it was called Lake Bussa by some of the local people, Lake Rudolf by Count Samuel Teleki, who "discovered" it in 1888, and Lake Turkana, the name given by the Kenyan government after independence in 1963, in recognition of the Turkana people who live on its western shore.

The lake, shaped like the crook of a dog's leg, measures 180 miles from north to south, with an average width of twenty-five miles. It is a formidable piece of water, a powerful presence for the people who live near it and even for those who visit it only briefly, like the scientists who work there. I don't know anyone who has spent time at Lake Turkana who doesn't feel that it is in some way home. Strange, for such an inhospitable environment. It is home to me in many important ways.

Our landing strip is two thirds of the way up on the western shore, seventy-five miles north of Lothagam Hill, so we had about half an hour more to go. As the Cessna's nose kept its northerly bearing, I could already see, far up on the eastern shore, the familiar spit stretching out into the lake. That's the Koobi Fora spit, home base for a decade and a half's fossil hunting that put me precisely where I had vowed I would never be: following in my parents' footsteps. But not, I think, in their shadow.

I cannot explain—even to myself—how I was drawn into research on human origins, following a path I'd so strongly vowed to avoid. It was partly accident, as these things so often are. As a youth I was a good organizer—I knew that. And I could run expeditions in the kind of difficult terrain one often encounters in the pursuit of fossils in East Africa. Slowly but inexorably, I became involved in running the practical side of such expeditions. And slowly but inexorably I became fascinated by the fossils themselves. If I'd known then what bitter academic and personal battles lay ahead, maybe I would have dropped the whole enterprise and gone off to do something more peaceful—like being an army general. But fossil hunting was in my blood, and I couldn't escape it.

One experiences a powerful sense of awe in holding a hominid fossil, a piece of one's past, a piece of the past of all of Homo sapiens. It never fails to thrill me, and I know I'm not the only one in this profession to react this way. My colleagues and I don't talk about it much, because it doesn't sound scientific, but it's a very real part of this very special science, the search for the identity of all mankind. Maybe that's what drew me on.

My decision to explore the fossil potential of Lake Turkana's eastern shore had been a wild gamble, the kind you take when the arrogance of youth blinds you to the likelihood that you will almost certainly lose. A thunderstorm led me to the decision.

The time was August 1967, and I was leader of the Kenyan team in a joint French-American-Kenyan expedition to the southern Omo Valley, just north of Lake Turkana. My father had been instrumental in setting up the expedition, through his friendship with Emperor Haile Selassie, the Ethiopian leader. I was therefore very much aware of Louis's presence, even though he was not part of the venture.

After surviving, early on, the best efforts of a giant crocodile to consume the Kenyan team and the flimsy boat that carried us across the giant Omo River, we had a bit of luck. We found fragments of two relatively recent human skulls, each about 100,000 years old, each an example of an early modern human, that have subsequently been recognized as evidence in the human story. However, at the time we were all much more interested in the earlier part of human history, and this piece of luck just wasn't big enough to satisfy me. I also felt that the Kenyan team had a poor deal in the expedition. Each team was allocated discrete geographical regions in which to operate, and the majority of the fossils in my area were clearly much younger than those in the other localities. It was clear to me that we were likely to be eclipsed by any discoveries that the French and the Americans might make. I couldn't bear the thought that the paleontological prizes were going to be scooped up by the other teams.

As it turned out, not much of great import was found by those teams, except an unprepossessing human fossil, a 2.6-million-year-old lower jaw that its French discoverers called *Paraustralopithecus aethiopicus*. Almost two decades later this little jaw would play another role in my life, but at the time of its discovery I was not much interested in it. I was too concerned with the poor prospects of my team and, perhaps even more, with my

17

own status. Yes, I was leader of the Kenyan team, but I had no scientific credentials, no formal education. I was a good organizer, but all I knew about anatomy was what I had learned from my boyhood business of selling skeletons to museums and what I had picked up by working alongside scientific colleagues. Effectively, my father was the scientific leader of the Kenya team, and that irked me.

Toward the end of the Omo expedition, I had been flown back to Nairobi to take care of some business. During our return, as we neared our destination, a huge thunderstorm roiled over the western edge of the lake, forcing the pilot of our small plane to take an easterly route. I was familiar with maps of the area, which showed the eastern shore to be covered by nothing but volcanic rock, so I was surprised to see below me what appeared to be stratified sediments, just the sort of formation that might contain fossils. I knew that, for a variety of reasons, no one had surveyed the area in search of fossils. So I determined to look myself.

A few days later, in a helicopter hired by the American contingent of the expedition, I was being flown over the same northern reaches of the eastern shore I'd seen from the plane. I asked the pilot to put down close to some likely looking sediments, and within minutes I was holding fossils and stone tools in my hands. We explored several other sites that day, and I began to see the future. I knew immediately what I had to do, but I kept my plans secret.

Four months later, in the grand board room at the headquarters of the National Geographic Society in Washington, D.C., I reported on the progress of the Omo expedition. Then I set out my proposal for an exploratory expedition to Koobi Fora, on the eastern shore of Lake Turkana. I described my brief helicopter visit and told the committee what I had found. I was confident, I said, that many fossils were there. The cost would be $25,000.

My father was shocked. Although he knew of my interest in investigating the Lake Turkana area one day, he assumed I was at the meeting to ask for funding for further support of the Kenyan part of the Omo project. The committee was taken by surprise too, not least at the audacity of an uneducated twenty-three-year-old seeking generous financial backing for an independent expedition. I had left high school early, eager to pursue my own interests in the world. I had no university education, nor, I think, the required patience for any. And yet here I was, seeking research support that might otherwise have gone to a "real" scientist. In spite of my bluff, I really did not know what would come of an east-side expedition. I just knew I had to try. The National Geographic decided to back my gamble.

Alan Walker and I were now level with the southern end of Turkana, directly to our right. Ahead, the lake stretched on and on, an endless glistening sheen, eventually melding with the morning mist. Nairobi was truly far behind; I felt released from the pressures of the city, from the demands of the museum. The surge of tranquillity never fails when I reach this point in the journey. I looked again to my right and saw South Island apparently gliding slowly by, and I was reminded of one reason that the fossil-rich deposits of the eastern shore were unexplored until I led my first expedition there in 1968. It concerns the death of two young men.

The British geologist and explorer Vivian Fuchs had organized an expedition to the lake in 1934, with ambitious plans for extensive geological, paleontological, and archeological surveys. "The original plan for the expedition was that it should make a continuous journey round the lake," Fuchs told a gathering in the grand surroundings of the Royal Geographical Society in London on the evening of April 15, 1935. "Owing to the refusal of the Ethiopian government to grant permission for the expedition

19

to enter Abyssinian [Ethiopian] territory, this plan had to be modified so as to omit the extreme north end of the lake, which lies just over the border. It was therefore decided to conduct the work in two sections, first on the west side of the lake and then on the east." Precisely the two areas where, four decades later, my work would be done.

The west side proved disappointing to the Fuchs expedition, as Mr. D. G. MacInness, one of Fuchs's associates, explained to the assembled eminent scientists and explorers at the Geographical Society. "We knew of certain fossil deposits of supposedly Miocene Age on the west side of the lake, where fossils had been obtained by the French expedition about two years before," said MacInness. "We did actually find some of the diggings made by the French, but either they had collected everything, or there had not been a great deal to collect. Practically nothing was to be found." So the expedition turned its attention to the eastern shore.

One of the explorers' interests was South Island, an extinct volcano about four miles from the eastern shore and fifteen miles from the southern end of the lake, which has been variously called Hohnel Island, after Lieutenant Ludwig von Hohnel, who came across the island during Count Teleki's 1888 expedition, South Island, and, incorrectly, Elmolo Island. El Molo is the name of a people who live on the eastern side of the lake. There is an Elmolo Island, but it is farther to the north and is much smaller than South Island. Whatever name is applied, the island has always been the subject of myths and legends among the people of Lake Turkana. They talk of fires that were seen there long ago. Since the island is of volcanic origin, this seems very reasonable.

On July 25, 1934, Fuchs was accompanied by Mr. W. R. H. Martin, a surveyor, on a visit to the island after a short but difficult sail. The following day the two men explored parts of the island, heading for the highest peak, fifteen hundred feet.

"After a while we saw the tracks of some four-footed animal: a surprise indeed, for we had supposed the island to be uninhabited except by birds," recounted Fuchs. The tracks turned out to be those of goats, domestic goats run wild. "Later we found thirteen goat skeletons in various parts of the island, and also a fragment of broken pot and some human bone." So maybe this was the source of the fires that had been seen on the island, not volcanic vents, after all. In any case, the discovery was a morbid prelude to another mystery of the island.

On July 28, Fuchs returned to the mainland, leaving Martin to continue his surveying work. The following day, Dr. W. S. Dyson, an American and the expedition's medical officer, joined Martin for what was to be about two weeks of difficult but fascinating scientific work. By this time the expedition had completed its explorations of the western side of the lake, and the excursion to South Island was part of similar work on the eastern shore, where fossil and archeological discoveries were confidently expected.

Elaborate plans were laid for emergency communication between the island and mainland camps, but none was ever made. The time for rendezvous came and went without any kind of signal at all. Whatever happened to the two young men must have overcome them quickly, for they disappeared without any call for help. An intensive and frustrating search, involving several aircraft and boats, failed to locate them. All that was ever found of the men and their enterprise were two tins, two oars, and Dyson's hat, all of which washed up on the western shore about seventy miles north of South Island. "Not the least mysterious aspect of the whole affair was the disappearance of the boat and the two four-gallon buoyancy drums that were carried in it," said Fuchs.

Exactly forty years later, in the summer of 1974, a tragedy struck an expedition of mine, just a little farther north on the eastern shore of the lake.

A young graduate student was out collecting samples on his own—a dangerous contravention of camp rules. He became lost. Again, a massive aerial search over a period of four days failed to locate him. He was found in the end by pure chance, but he was so ravaged by heat and dehydration that he was delirious, and we never discovered what happened. He died a few days later in a Nairobi hospital. As with the case of Martin and Dyson, the loss of this young student was a grim reminder that Lake Turkana, the place I love, can be harsh and unforgiving, and demands respect—like all of nature.

As far as can be discerned from the records and word of mouth, Fuchs and his young colleagues had reached a point just south of Alia Bay when the deaths of Martin and Dyson brought an end to their expedition. The expedition's stated purpose—"finding remains of early human cultures"—went unfulfilled. But the men were close, because the area they reached is one I now know well—precisely where the fossil-bearing strata start to become interesting. Sometimes I smile a little when I think how close they came.

We had begun our descent toward Lothagam Hill, and the temperature in the cockpit was already climbing. Soon, the air would take on the familiar smell of Lake Turkana, an odd mix of burned grass and parched earth. Alan was scanning an area called Karukongar, anxious to spot the tuff, the layer of volcanic ash that here has an odd bluish color. "Which way will you take it, Leakey?"

Lothagam Hill is what geologists call a horst, a vast slab of rock heaved up at an angle toward the western edge of the rift. What caused the slab to lift like a giant trap door, we don't know, but the effect is like the structure of the Basin and Range of the American West. Here, the horst is an island amid the flat, narrow floodplain west of the lake. Just ten thousand years ago

Lothagam really was an island, because Lake Turkana then was massive, submerging an area at least four times that of the present lake. Not long ago burial or ceremonial sites of about the same age—ten thousand years—were discovered in the saddle of the hill. It is intriguing to think about a small community on Lothagam Island.

A combination of volcanic activity, sedimentation, and time —about four million years—transformed Lothagam into a paleontologist's dream and a geologist's nightmare. There are pockets of sediments rich in fossils, and in one small gully an entire skeleton of a carnivore is slowly eroding out of the sandstone wall. But elements of the geology are difficult to untangle, which complicates the business of dating the fossils we find there. So it is with the Lothagam *Australopithecus* mandible, the piece of fossilized lower jaw that I mentioned earlier. Its age remains an enigma.

"Look, there it is," said Alan. To our left was the bluish tuff, from this height looking like a broken streak across the landscape. On the ground it is much more impressive, in places measuring ten feet thick, a river channel in stone. "Let's see if we can trace it to Lothagam," I replied as I put the nose down, ready for a little bush flying. Unfortunately, the tuff proved to be too fragmentary, and, try as we might, we could not trace its path. We gave up and decided to look again from the ground, where we would collect some samples, for chemical analysis, in a day or two.

We had found the tuff three years earlier, when Alan and I did a safari to Karukongar with Kamoya and the Hominid Gang, working our way up from just north of Lake Baringo to this spot just south of Lothagam. It was a quick trip, a survey for interesting prospects, made memorable in part because Alan's arm swelled up to alarming proportions after he was stung by a wasp. Being allergic to wasp venom, and therefore likely to go into life-threatening shock, makes for interesting times when one is out in

23

the bush, hours away from any hospital. He recovered soon, however.

Our final camp on that trip was about ten miles south of Lothagam, by a big sand river lined with trees that gave welcome shade. But we weren't alone, as is often true on the relatively populated western shore of the lake. We knew there were herders around, mostly because of their animals, hundreds of camels and goats. Some old Turkana men came into camp for tea and tobacco, but there were no young men to be seen. We should have been more suspicious than we were. Soon afterward Alan and I had to fly back to Nairobi, leaving Kamoya and the Hominid Gang to pack up camp the next day. As it happened, they had to pack up camp rather earlier than planned.

As soon as the sound of the airplane died away, the missing young men appeared, carrying rifles and demanding blankets and other camp items. Somehow Kamoya managed to convince the men—local bandits, it later transpired—to come back the following day, when everything they wanted would be handed over. After the bandits left, Kamoya had his men secretly pack up everything they could inside their tents so that it wouldn't be obvious that they were abandoning camp. Then, on a signal about four hours after sunset, guide ropes were cut, the tents quickly thrown into the Land Rovers, and the party fled at speed into the dark night, with no headlights. "It was a little difficult," Kamoya later told me in his understated way.

Satisfied that we'd learned all we could about the puzzle of Lothagam from our aerial survey, Alan and I decided to make for camp. Kamoya would be waiting. The airstrip is short. I prefer it that way, as it discourages unwanted visitors. With the lake behind us and the western wall of the rift ahead, we made a steep descent, and the temperature continued to rise. Some people are nervous about bush landings, especially mine. But I don't take risks—not anymore, anyway. There was a time—I was much younger—when I thought I could do anything and get away with it. A narrow escape taught me I was wrong. I prefer to go on living.

The altimeter showed fifteen hundred feet. The stall warning buzzed urgently. The tires hit and then rumbled over the parched earth as I braked hard. I flung open the window, and a blast of hot Turkana air enveloped us, carrying the scent of distant goat herds and parched vegetation. We saw a Land Rover, and Kamoya and Peter Nzube waiting, smiling broadly. "OK, Walker," I said to Alan, "let's see what they got for us this time."

2

A Giant Lake

"*K a m o y a   h a s* found a small piece of hominid frontal, about 1.5 by 2 inches, in good condition," Alan recorded in his field diary on August 23, 1984. "It was on a slope on the bank opposite the camp. The slope itself is covered with black lava pebbles. How he found it, I'll never know."

Kamoya's skill at finding hominid fossils is legendary. A fossil hunter needs sharp eyes and a keen search image, a mental template that subconsciously evaluates everything he sees in his search for telltale clues. A kind of mental radar works even if he isn't concentrating hard. A fossil mollusk expert has a mollusk search image. A fossil antelope expert has an antelope search image. Kamoya is a fossil hominid expert, and there is no one better at finding fossilized remains of our ancestors. Yet even when one has a good internal radar, the search is incredibly more difficult than it sounds. Not only are the fossils often the same color as the rocks among which they are found, so they blend in with the background; they are also usually broken into odd-shaped fragments. The search image has to accommodate this complication.

In our business, we don't expect to find a whole skull lying on the surface, staring up at us. The typical find is a small piece of petrified bone. The fossil hunter's search image therefore has to have an infinite number of dimensions, matching every conceivable angle of every shape of fragment of every bone in the human body. Often Kamoya can spot a hominid fossil fragment on a rock-strewn sediment slope from a dozen paces; someone else on his hands and knees staring right at it might fail to see it.

I met Kamoya in 1964, on my first serious foray into the hominid fossil business. He was part of a team of workers on an expedition to Lake Natron, just over the southwest border with Tanzania. We immediately struck up a friendship and professional relationship that has continued ever since. He demonstrated his skill even then, by finding a fossil hominid jaw of the same species as *Zinjanthropus*, which my mother had discovered five years earlier at Olduvai Gorge. Kamoya's find was the only known lower jaw of this hominid species, so I was very impressed. Particularly so as Kamoya spotted the fossil barely protruding from a cliff face not two feet from where I had been looking for fossils a little earlier. Part of Kamoya's secret is that, although he's a stockily built man with a great sense of calm about him, he is always on the move, restless, rarely idle. So it was when he found the piece of hominid skull that had brought Alan and me on our trip to west Turkana.

"We had our camp by the Nariokotome River," explains Kamoya. "It's dry most of the time, but about a hundred yards upstream from the camp you can dig down and find water, two feet down if there has been rain recently, maybe ten feet if it's been very dry. But you can always find water." Kamoya and his team were on their way from the northern part of the western shore to some areas in the south, where we knew there were some promising fossil deposits. The geologist Frank Brown and the paleontologist John Harris were part of this north-to-south sweep, the final stages of a four-year survey of likely fossil localities on the west side. We had decided that 1984 would be the year serious work began in the search for hominid fossils there. And we had reason to be optimistic, because a couple of small fragments had been discovered early in the survey.

Nariokotome had been the site of a camp the previous year, so Kamoya knew that shade and water could be found there. "We arrived about midday, dirty and tired," he recalls. "The first thing we did was to look for water. Yes, it was there, just like last year, except we had to dig a little deeper this year." Their bodies and clothes washed, lunch eaten, the men declared the rest of the day was a holiday. Not for Kamoya. He thought he would take a look at a gully across the dry riverbed, just three hundred yards away.

"I don't know what it was about that gully that attracted Kamoya to it," says Frank Brown, who had been with Kamoya the previous three seasons of the west-side survey. "We passed by in 1981, the second year of the survey, and he took a look then but found nothing. It was the same the next year. Nothing. And then this year, 1984, bingo! He finds a hominid." Kamoya's explanation is typically enigmatic: "It just looked interesting." I count myself a fairly skilled hunter of fossil hominids too, and I occasionally get a sense—nothing tangible—that I'm going to find something, so I understand what Kamoya means. But even to my eye this gully looked unpromising, a scatter of pebbles on a slope, a goat track snaking by a ragged thorn bush, the dry bed of a stream that cuts the gully, and a local dirt road just a few yards away, running north to south.

"The soil's a light color in the gully," explains Kamoya, "and the stones are black, pieces of lava. The fossil is a little lighter than the lava, so it was easy to see. I found what I was looking for." The fragment was not much bigger than a couple of postage stamps put together, but nevertheless it was diagnostic. A flattish piece of bone with a slight curvature indicated skull, and a skull from a big-brained animal. In addition, the impression of the brain on the inner surface was very faint. Together, these clues triggered Kamoya's search image to say *hominid skull*. A similar piece of bone, thinner, with a tighter curvature and with deeper brain impressions on the inner surface might have indicated an antelope, for instance.

It wasn't immediately obvious where on the hominid skull Kamoya's fossil fragment had come from, but it turned out to be a part of the frontal region. Kamoya did know that the skull was more than a million years old—1.6 million years, according to Frank Brown's calculation—so he guessed he had found a *Homo erectus*, the hominid species directly ancestral to *Homo sapiens*.

The earliest member of the hominid family evolved somewhere between five and ten million years ago, according to current estimates. A good average date, therefore, is 7.5 million years ago for the origin of the first hominid species. One of the defining characteristics of hominids is the mode of locomotion: we and all our immediate ancestors walked erect on two legs, or bipedally. Although the earliest members of the family were bipedal, and therefore had their hands free from the immediate business of locomotion, the making of stone tools and the expansion of the brain came relatively late in our history, beginning about 2.5 million years ago. There is some debate about it, but I am convinced that the making of stone tools is a characteristic of our own branch of the human family, the *Homo* lineage, and that it is closely associated with the expansion of the brain. The evolutionary increments in these respects were small at first but became significant with the appearance of *Homo erectus*. As we shall see throughout this book, the origin of *Homo erectus* represents a major turning point in human history. From the vantage point of today, it speaks of leaving an essentially apelike past and embarking on a distinctly humanlike future. For this reason, Kamoya's discovery was potentially very important.

"I called my people over," says Kamoya, "and we searched the ground surface. We found one more piece, but that was all. So we built a pile of stones, a cairn, to mark exactly where the fossils had come from." It was too late that evening to call me in Nairobi, so Kamoya had waited until the following morning to give the news. In fact, the news was twofold, because a little earlier John Harris also had found a piece of skull, which he thought might be a hominid or a large monkey. This one was about two million years old, again according to Frank's initial estimate. So when Kamoya and Peter met Alan and me at the airstrip, there was a lot to talk about, plans to make for further exploration of the two fossil discoveries—and, of course, camp gossip.

The airstrip is on the south side of the Nariokotome River, the camp on the north side, shaded beneath tall *Acacia tortilis* trees along the soft, sandy bank. A scatter of green canvas tents, the camp is a simple arrangement whose central focus is the mess tent where we eat, study fossils, and talk. The lake, three miles to the east, is invisible to us there, but we can feel the breeze that blows frequently from east to west across the Turkana basin. Grass grows only in patches, and the mostly barren landscape is here and there broken by the bushy shape of salvadora trees, their light green and luxuriant foliage obscuring the abundance of tiny, peppery fruit that the primates hereabouts—human and nonhuman—greatly enjoy. Acacias of various types line the river courses, which often cut deep gullies in the ancient sediments. The western shore of the lake has far more vegetation than the east, especially along the rivers, and there would be even more but for the heavy grazing by the Turkana people's goat herds and a few skinny cattle. The mountains impose their presence palpably here; the western wall of the Rift Valley is silhouetted against the sky. It is a magical land, trapped between a giant lake and majestic mountain ranges.

Like high mountains, great lakes have always drawn people

to them, explorers who need a target of discovery, an accomplishment to fulfill personal goals or to make them famous. In the literature of European explorers of the late nineteenth century, Lake Turkana, then called Lake Rudolf, figures prominently, often as the goal of major expeditions. "Again and again you come across references to the way the lake level has 'dropped dramatically' or 'risen dramatically' in just a few short years," says Frank, who has an epic collection of historical accounts of the region. "It's not surprising, then, that the image of this great body of water in a constant state of fluctuation is the one that came to dominate our views." Not surprising at all, because I myself have seen the level drop more than thirty feet in the past twenty years.

The magnitude of the basin and the scale of the water flow are barely comprehensible. Frank calculates that, with eight thousand gallons a second pouring in from the Omo, the lake could fill from scratch to its present level in just seventy years. That means the level is bound to be unstable, and just slight alterations in gain or loss have a big impact. Nevertheless, it is hard to grasp that at times there was no lake at all. But we know that it's true.

The realization began to dawn in the early 1980s, when Frank came to work first on the east side of the lake and then over on the west side. Previously he had been part of an American team, led by Clark Howell, that was working in the lower Omo Valley, a continuation of part of the American-French-Kenya joint expedition I had abandoned in 1967. Frank's job was to assemble an account of the region's geological history, a record of the environmental changes entombed in deep sediments.

Episodes of higher-than-normal rainfall, for instance, or periods of aridity, leave clues in the sediments that accumulate through time. The presence of a lake, the floodplain of a nearby river, these too impress themselves on the geological record. Wherever we have sediments, we can reach back into the past to read a record of vanished environments and the changes they experienced. A time scale for these changes is provided by inter-leafing layers of volcanic ash, known as tuffs. Radioactive iso-topes that are a natural component of the volcanic ash enable us to pinpoint the date of the eruption that produced the ash, be-cause the steady decay of the isotopes acts like an atomic clock. The so-called potassium-argon dating is one of the most com-monly used methods for creating the record of eruption events in East Africa. Frank had meticulously reconstructed such a rec-ord for the lower Omo Valley, and he then did the same for the sediments east and west of Lake Turkana.

"I began to realize that there were intervals when there was no lake in the basin," explains Frank, recalling his earlier work. "The closer I looked at the data, the clearer that became." But Lake Turkana is so big, has so much presence, dominates our lives here so much, it is difficult to imagine a time when it wasn't there. It seems unthinkable.

The lake still presents us with enigmas, including times when the vast Omo failed to empty into the basin at all. Frank guesses that, for some reason, the waters of the Omo were temporarily diverted and flowed toward the Nile. Someday we may know every slice of the region's history. But the important thing is that we know enough to accept that the things we see around us today are perhaps just brief moments in history, not necessarily an accurate guide to how things were in the past, or, indeed, how they will be. If we wish to gain a perspective of human history, as I fervently do, that is an important lesson.

In recent years I have come to believe that this perspective is perhaps the most important lesson to be learned about ourselves. *Homo sapiens* occupies the slenderest of time slices in Earth his-tory, a brief, passing moment. Our planet is some 4.5 billion years old. Primitive life here began almost four billion years ago; the first life forms on land appeared some 350 million years ago; the first mammals, 200 million years ago; the first primates, a little more than sixty-five million years ago; the first apes, thirty million years ago; the first hominids, about 7.5 million years ago; *Homo sapiens*, perhaps 0.1 million years ago. There is so much in Earth history to enthrall us, so much complexity and richness to thrill us, and yet, ineluctably, we are drawn to focus most intently on our own origins.

This passion to know, to know how we came to be and what made us what we are, surely reveals to us something about our nature. We are creatures of knowledge, it is true. But, more important, we are creatures driven to know. This passion to know is what brings me and my colleagues to the shores of this ancient lake basin, where for four million years our ancestors were part of the world of nature, a drama shaped by the forces of chance and natural selection. Like Count Teleki, we were drawn to Lake Turkana for a journey of discovery. But, unlike the count, we see that the object of our discovery is not the lake but ourselves.

"We have many bones to show you," promised Kamoya as we unpacked the belly of the plane. "You will like the hominids." I knew I would. "Skeletons?" I joked, and we all laughed at the improbable prospect. With evening upon us, we drank beer by the mess tent; the darkness fell quickly, as it always does this close to the equator.

Over dinner we discussed our plans, the thought of new hominid fossils uppermost in our minds. I proposed that our first visit the next day be to John's site to see the fragment that might be hominid or a large monkey. If the fossil really was hominid, and if it really was two million years old, it could be very important. The hominid story around two million years ago is unclear but is crucial to the origins of the big-brained creatures that eventually became us, so any new fossil is potentially illuminating. I had the feeling that one day we would be in for some real surprises in this slice of our prehistory, two million years ago.

Perhaps John's fossil would provide it. On the other hand, I was not optimistic about the prospects of Kamoya's *Homo erectus* site. "Seldom have I seen anything less hopeful," I recorded in my diary before turning in that night, tired but happy to be at the lake.

Up the next morning at five-thirty; tea, bread, and cheese for breakfast; and we were on our way by first light, about six o'clock. It was a slow drive, the twelve miles south to John's site, on Laga Kangaki, another dry riverbed. Strong shadows from the low-angled sun, cool, sweet air: it was a delightful Turkana morning, heightened by a sense of anticipation about what the fossil site might yield.

While I drove, John pointed out places where he'd found other fossils. John's interest is principally in vertebrates other than hominids. His role in our expedition was to collect information about the ecological community that occupied the region between four million and one million years ago. While hominids were a part of that community, John wanted to construct a picture of the ancient environment that included antelopes, zebras, elephants, giraffes, hippos, pigs, hyenas, and various primates. This kind of information is crucial to an understanding of the ecological setting in which our ancestors lived and how it changed with time.

Finally we arrived, and John led us to the site, a small slope with a few fossils, typical of west Turkana. Like Kamoya, John had marked the location of the fossil with a cairn. The fragment was about the same size as Kamoya's find, and, by chance, came from the same part of the skull, the frontal region. And, yes, it was a hominid. "Maybe this is your big find, John," I said. John laughed nervously. "You said you wanted something special," he responded.

When a hominid fossil is found, an electricity of anticipation tingles down the spine. We know from experience that most finds are single fragments, a piece of skull or another part of the anatomy, and nothing else will turn up. That's the nature of the fossil record—woefully incomplete, as Darwin observed. But we also know that there is a chance that the next find, the next fragment, may be the beginning of a major discovery. "Well, let's take a look," I said as we carefully crouched near the cairn, scanning the surrounding terrain for other pieces.

I pointed to the edges of the fragment. "Fresh breaks." Often, when an animal dies in the kind of terrain that has existed for millions of years around Lake Turkana, its skeleton is trampled and broken by passing herds. The fragments may then be buried and slowly petrified, to become part of the fossil record, and the edges of the fragments would be fossilized as ancient fractures. If, however, a complete cranium is buried and becomes fossilized, we see something different. Best of all, as erosion slowly exposes the cranium millions of years later, it remains intact—to be picked up by a lucky fossil hunter. A rare event. Much more likely, when the cranium is slowly exposed by the elements, it breaks into fragments of various sizes. But the breaks—the edges of the pieces—will be fresh, indicating that the rest of the cranium may be scattered in the vicinity. That is what we saw in John's hominid fragment. "Looks as if we may be on to something," I said, and our excitement built as we began to search for what might be an important find.

Unfortunately, our initial search turned up nothing, so we decided to sieve—later. Sieving is done by carefully scooping up the loose material on the surface where the fossil has been found and then passing it slowly through mesh. This separates the tiny grains of soil from pieces of stone and, we hope, fossil. It is a time-consuming and tedious business, and we all try to find urgent business elsewhere when sieving time comes. It had to be done, but not just yet.

With sieving operations to be organized later, we spread out over the landscape, looking for other fossils. John found a skull of a crocodile, and Alan and I excavated a baboon skull and later a bovid skull, some kind of gnu, or wildebeest. The gnu was a gem of a fossil, but very fragile and broken. We applied Bedacryl, a kind of glue that hardens fragile fossils, and planned to pick up the skull in a few hours after the glue had set.

By now the sun was high and it was becoming very hot, so we returned to camp. Over lunch we decided that the rest of the Hominid Gang would start sieving operations at Kamoya's site, which was close to camp. Alan, John, and I would join them. "We sieved for about two hours," Alan recorded in his field diary that evening, "picking off the boulder lag and screening. It is very dusty and the stones are black." It wasn't pleasant, and I knew it would get worse.

Nevertheless, we chattered and joked and were entertained by the antics of a band of young Turkana children swarming over a huge salvadora tree just opposite the site. The tree was in fruit, so the smaller children were able to clamber among the branches and eat well. The berries taste a little like nasturtium leaves. Quivering with the movement of the children hidden among the foliage, the tree seemed animated or possessed, especially when we heard the high-pitched giggles of pranksters.

After two hours of scooping dry earth, shaking it through mesh, and finding absolutely nothing of any interest, we found our enthusiasm waning, and Frank asked whether we'd like to see some fossil stromatolites he'd found. Needing no more of a pretext, Alan and I excused ourselves from the sieving and set off. John came too. We all thought nothing more would come of the work at hand.

Stromatolites are like primordial beasts, "like a herd of hippos wallowing in the mud," as Alan puts it. In fact they are colonies of single-celled algae and other microorganisms that grow in shallow, calm water. The colonies begin life very small, centered perhaps on a grain of sand, and grow by adding layer upon layer upon layer, producing a colony the shape of a flattened sphere. "Beautiful things," says Frank, "like the caps of

giant mushrooms." Some of them can grow to three or four feet
in diameter, the size of a small table. Although they are rare in
today's world—Shark Bay, Western Australia, is one of the few
places where living examples can be seen—stromatolites are
common in the fossil record, and are among the first signs of life
on Earth. For instance, fossils of some simple forms of stromatolites date back an astonishing 3.5 billion years.

We drove westward, away from the lake, about a kilometer,
and up to a point close to the riverbed adjacent to the Nariokotome. Soon we came across Frank's stromatolite field. The
fossils were only about a million years old, mere youngsters in
the grand scheme of things, but impressive enough. I had not
seen them before, lined up like a causeway on a plain below a
ridge strewn with lava boulders. The plain itself was virtually
barren, save for a scattering of small gray-green plants. There
was a covering of fine volcanic gravel. It looked like a Japanese
garden. Some of the stromatolites were split, and we could see
the successive layers as the colony grew, like tree rings. Quite
apart from their fascination, the "creatures" are proof that a million years ago the lake edge was right here, where today it is arid
desert, the lake shore three miles distant. When we were back on
the ridge, I turned, looked in the direction of the lake, and tried
to imagine what it would have been like a million years ago at
the ancient lake edge, with living stromatolites at my feet.

The shallow lagoon before me is one of many that interfinger the
floodplain sediments of the western shore, part of the restricted
lake system of the time. Reeds and other aquatic plants that can
tolerate the brackish waters fringe the lagoons. A spiky grass
covers much of the floodplain, a thin, pale green carpet for much
of the year. When the seasonal rains come, however, tiny flowers
spring up among the grass, a purple fringe near the shore giving
way to a yellow border farther from the lake. Even the thorny

acacia bushes are covered with white blossoms, a cloud of tiny flowers that unfurl just before the rains come.

To the south and to the north of the lagoon, seasonal river courses squirm their way from the mountains down to the lake, edged by acacia woodland with a scattering of fig trees. Near the lake, these narrow refuges give way here and there to sedges and marshland. Between the green-fringed river courses are regions of semi-arid scrub, a reminder of the unforgiving nature of this land, a reminder that without the inflow from the majestic Omo River, all before me would be hot, arid, and barren, a basin of death rather than of life.

Beneficiaries of this extraordinary oasis abound, from elegant colobus monkeys feeding on the figs in the gallery woodlands, to three different species of hippo wallowing in the lake; from the giant warthog rooting among the bushes to a curious giraffelike creature, with a short neck and enormous horns. Down by the shore countless birds feed on the bounty of the lake, some fishing, others filtering the nutrient-rich waters, and still others perching on the curiously long snouts of crocodiles, scavenging what they can. On the floodplain, herds of zebra and gazelle graze the sparse grasses, and donkey-sized three-toed horses are playing out the end of their evolutionary existence.

And predators lurk, waiting their chance to strike. Two species of leopard, jackal, hyenas, and saber-toothed cats, another species nearing the end of its life.

From a distance, it all looks much like today's community: the lake and its tributaries supporting a range of plant life; grazers and browsers foraging freely amid a parched and arid desert; and predators taking what they can. It is an interplay of forms of life of the moment, something we recognize as very familiar. But not completely familiar. The saber-toothed cat and the oddly formed giraffe are no longer here; they are victims of extinction, as is the three-toed horse. Even the species we more readily recognize, like the gazelle, the wildebeest, the pig, even the zebra, all creatures of today's African landscape, are different in various ways, often subtle. Differences in the size of body, the shape of horns, the configuration of head—all evolutionary variations on persistent themes.

But there is something entirely novel out there too, at least from the perspective of humans viewing an African landscape. Troops of large primates forage in the woodland and out in the scrub, characteristically noisy, but uncharacteristically moving around on two legs. Large, bipedal primates, like us. One troop keeps returning to a sandy bank near a river course, as if the location served the purpose of a campsite. They are tall, muscular, and powerful, these primates, and can run smoothly and with persistence. Some bring plant foods to the river-site camp; others carry pieces of carcass, the remains of a small antelope; still others transport rocks, retrieved from a distant outcrop. The noise of their constant communication is now joined by the sound of rock striking rock, flakes expertly being detached. One individual uses the flakes to strip bark from a branch; another uses them as a butcher's knife. So much activity, so very familiar, and yet uncannily different.

A few miles away, among the woodland strip of another river course, another troop of bipedal primates is noisily foraging, some in the trees, others nearby, digging for tubers. But there is no obvious campsite here, no to-ing and fro-ing with pieces of carcass or lumps of rock. Yes, there is a lot of vocalization, typical primate communication, but it is somehow more restricted than among the first group, less richly patterned. They are bipedal; about that there is no question. But their trunks seem bulkier, their legs shorter, and they don't move with the power and ease of those in the first troop. They are clearly different animals, a different species of bipedal primate. Variations on an evolutionary theme.

These two types of bipedal primate are very much a part of the ecological community of the Lake Turkana basin, part of the interplay of the different forms of life there. But, while one of them is living like a typical primate, like a baboon, for instance, the other is clearly extending the boundaries of what it is to be a primate. The result is an emerging evolutionary novelty, whose future, a million years ago, could not have been predicted; whose future is us.

I was brought back from my reverie by calls from Frank and Alan, insisting we move on. I looked again at the stromatolites at my feet, once more a petrified instant from a vanished past. We visited several other fossil localities as John told us more of what he'd found in the weeks before we arrived, more of our ancestors' environmental setting. We headed back to camp, giving little thought to the sieving task we had left behind at Kamoya's hominid site. But as we neared the shade of the Nariokotome camp we heard people shouting: "We've found more bone! Lots of skull!"

We ran to where Kamoya was sitting, his treasure arrayed before him, like jewels plucked from the dry earth. "The right temporal, left and right parietals, and bits of frontals of a beautifully preserved (if broken) *Homo erectus*," is how Alan described the find in his field diary. "That's a lesson," I later noted in mine. "The most unpromising site, as Kamoya's surely was, can sometimes surprise us." Like everyone else, I was elated. There was great excitement, joking, and laughter. Here, beginning to take shape before our eyes, was part of the front and sides of the cranium of a human ancestor, *Homo erectus*, upright man.

3

The Turkana Boy

A s k e l e t o n ? Could we really be on to a skeleton? It was August 30, 1984, exactly a week since Alan and I had arrived at the Nariokotome camp, but there had been only two full days of full-scale excavation at Kamoya's *Homo erectus* site. We hardly dared voice the speculation—the hope—out loud. As much as anyone, paleontologists believe in the dangers of tempting fate. And yet we already had a tantalizing haul: a lot of the cranium, part of the upper jaw and face, a fragment of cheekbone, some rib pieces, a small part of a shoulder blade, a vertebra, and a segment of pelvis. Yes, we might have a skeleton, but we knew it would take time and a lot of effort to find out. A major excavation was going to be necessary.

Two days earlier, I had returned from a visit to the east side of the lake, bringing my wife and daughters, Meave, Louise, and Samira, with me to Nariokotome. David Brill, a photographer for *National Geographic,* and two writers, Virginia Morell and Harriet Heyman, also arrived. The camp seemed crowded, and there was a sense of anticipation in the air, because tomorrow we planned to start serious excavation at Kamoya's site. Louise, who was twelve, two years older than her sister, was particularly excited; she was going to learn to drive a Land Rover this season. Samira agreed to help with work around the excavation, as long as we promised regularly to douse her with water to keep her cool. It was a deal.

While I was away from Nariokotome, Alan had been working with the material we had recovered in our initial search. "I realized that it would be made easier if we washed everything," he recalls. "Because of the fine dust, everything—the stone, the fossils—looked the same color, black. But after washing, the pieces of fossil showed up quite clearly. They were a beautiful mahogany brown." Alan rigged up the camp shower to do the job, carting the sieved material to it to be washed and sorted. Using the camp shower was to be a temporary measure, until we determined how much material we would have to handle. If the operation became big, we knew we would have to truck the stuff to the lake.

Alan had also begun trying to piece together the dozen or so skull fragments we'd found. There is a special skill in doing this kind of three-dimensional jigsaw. The pieces are the oddest of shapes, and there is no certainty that they will go together, because several parts of the puzzle are missing. Required for the task is a familiarity with the anatomy, of course, but especially a highly developed spatial sense. Alan has it, and so does Meave. With Alan, the talent also expresses itself in his sculpting, a talent I've noticed in a number of fine anatomists. With Meave, the spatial gift was evident when she was a young girl. She liked doing jigsaw puzzles, but found them too easy. So she would turn the pieces upside down, picture side facing the table, and still solve them.

Alan and Meave often worked side by side during expeditions at Lake Turkana, poring over fragmented fossils of one kind or another, slowly rebuilding shattered pieces of anatomy. "You can spend only so much time trying to get things to go together," says Meave, "and then you have to go away, do something else, let someone else have a go. Then, when you come back to it, you often seem to know exactly what to look for, as if your mind has been working on it while you're away. It's uncanny." On this occasion Alan had done a lot of the *Homo erectus* jigsaw before Meave arrived. "By six P.M. almost all the pieces we have are glued together," Alan noted in his diary. "The left parietal is nearly complete, the right less so, and the left frontal goes from midline and forward to nearly the posttoral sulcus." In other words, part of the top of the cranium, toward the front, was beginning to take shape, though there were still many gaps in the jigsaw. However, even if we found nothing else, we could be pleased with what we had. "We can set it up in the sand, and it's beginning to look like a real skull," Alan wrote that evening.

On the morning of the twenty-ninth, we set off early for Kamoya's site, everyone in tow. It is not often that an excavation site is just three hundred yards from the breakfast table, and it has its advantages. By the time we arrived, David Brill had already taken sightings of the area, testing the light, and was all set to take pictures. The work began, and we were soon rewarded. "We started excavating the rubble layer and found many pieces, including zygomatics [cheekbones], the other temporal, etc. And many more fragments which were recognizable," Alan wrote that night. "By lunch time we had begun to fill up a tray." In the shade of the camp, Alan and Meave occupied themselves with gluing the pieces together. And through the afternoon they had more and more to work with: part of the back of the skull, some rib pieces, a fragment of the underside of the cranium. "We were feeling pretty pleased by supper time," notes the diary.

And yet we had reservations. All the fragments were from the loose layer of soil on the surface, and there was a real chance that there was little more to be had. Perhaps an intact fossil

43

skeleton slowly eroded out of the ground a hundred years ago, two hundred years ago, who knows, and had since fragmented into pieces, leaving just the remnants we were finding. Judging from the distribution of the fragments on the slope, Alan and I agreed that this might be true. We tried not to think about it, but we were realistic.

The next morning we were back at the site and found nothing much, except a large yellow scorpion. We began to grow discouraged. After a few hours I pointed to a small thorn tree in the middle of the site and said, "Walker, if there's nothing more after this, we'll call it quits." The tree seemed to mark the level from which the bones had come, so it was a good stopping point. The tree was also a nuisance, as we kept snagging our clothes on its hooked thorns. (Not for nothing is it called the wait-a-bit thorn.) Then something extraordinary happened.

Peter Nzube was next to me, sweeping the excavated surface. I was working among the roots of the wait-a-bit thorn, gradually picking away at the ancient sediments that had given life to this young tree. Suddenly I saw something out of the corner of my eye and stopped. It was one of those uncanny instances when one's brain registers a split second before the mind is conscious: teeth, half an upper jaw, came the reflex message. Instead of using a brush to sweep the surface debris, Peter was using his hand. He had not expected anything important in the debris, nor had I, but there it was: a perfect half upper jaw, half a maxilla.

I was momentarily angry at Peter's uncharacteristic carelessness, but it was the excitement of the discovery that fueled the emotion. I punched him good-naturedly, and berated him heartily, to which he responded in the same vein. Our behavior startled Kamoya, because he hadn't seen what was taking place. The general commotion summoned the others, and we realized that we were looking at a sure sign that at least some of the fossil individual was entombed safely in the sediment. There might be more of the skeleton there; right beneath our feet might be the discovery of a lifetime. It was a breathtaking moment, a mix of realization and hope. We knew we wouldn't be quitting after all.

Alan and Meave were at the camp, gluing the fossil jigsaw together. Kamoya, Peter, and I took a break for coffee, calmed down, and extended the excavation. Samira trotted up from camp, holding a cleft stick with a note inserted in it, a caricature of the traditional way of carrying messages in Africa. "The gluing team got smart and counted teeth," read the note. "The third molar on the right side is missing. It's a sub-adult!"

Alan and Meave, working with the upper jaw, realized that the piece they were staring at was a diagnostic clue to the age of the individual. If all the molars had erupted, we would be dealing with an adult. If all the teeth had been deciduous, the set that is lost during maturation, then it would have been a child. But the permanent molars had started to come through, the first and second already in place, the third yet to appear. "This told us that the individual had died about the age of eleven or twelve," explains Meave. "Louise was twelve at the time, so we had a human model to compare the fossil with. She was also shedding her deciduous canines, as were some of her friends. So, it turned out, was the Turkana boy."

Our *Homo erectus* fossil was that of a boy, as we were able to discern from some characteristics of the pelvis. And we were soon in for a surprise on the rest of the boy's anatomy, something that turned out to be controversial when we announced the discovery later in the year.

During the next three weeks the excavation continued steadily, the skeleton becoming more complete with each passing day. It was an extraordinary experience for me, for all of us. Three weeks of paleontological bliss. When my mother visited us for a few days, she sat in the shade, making remarks about the scruffy state of our excavation compared with her precise digs at Olduvai Gorge. There is a running joke between paleontologists and archeologists, that the fossil people dig round, untidy holes, whereas the archeologists make neat, square-sided excavations, marked out by a grid of string. Like everyone else, however, she was deeply impressed with what she saw. "You have to go to Europe, to Neanderthal graves, to see fossil skeletons as complete as this," she remarked.

It was true. Of the many *Homo erectus* remains that have been discovered over the years, most have been parts of the skull; only a few pieces of the rest of the skeleton have been found at all. As a result, virtually every bone we turned up was the first of its kind to be seen by human eyes. "This is the first thoracic vertebra of *Homo erectus* known to science," Alan could be heard saying from the excavation. "This is the first lumbar vertebra of *Homo erectus* known to science," he would intone later. "This is the first clavicle of *Homo erectus* known to science." It became quite a litany, and we felt cheated when we didn't hear it.

Every day we took out fossils that in other years would have been sufficient to make whole expeditions worthwhile. We came to expect the bone haul to continue, and turned a little blasé. Even David Brill stopped taking pictures. "The first thoracic vertebra of *Homo erectus* you've seen, anyone's seen, and only the second pelvis ever seen—and you're not taking pictures," chided Alan. That night Alan wrote in his diary, "Just shows that the site is mind numbing."

Paleontological bliss, no question about it. Taking shape before our eyes was an essentially complete *Homo erectus* individual, the first to be unearthed since this species of ancestor was discovered almost a century ago.

Homo erectus stands at a pivotal point in human evolutionary history; in a very real way it is the harbinger of humanity. Everything earlier than *Homo erectus* was more apelike (except the short-lived, somewhat enigmatic *Homo habilis*). Everything after *Homo erectus* was distinctly humanlike, in behavior as well as form. The beginnings of a hunting-and-gathering way of life came with *Homo erectus;* stone tools for the first time gave the impression of standardization, the imposition of a mental template; fire was harnessed for the first time; for the first time hominids expanded beyond the African continent. And surely the rudiments of language—perhaps even consciousness—were produced in a dramatically expanding brain. Yes, *Homo erectus* marks a clear move from an apelike past to a humanlike future. If we are to understand the origin of humanity, we have to understand *Homo erectus,* its anatomy, its biology, its behavior. That's a tough challenge, given fragmentary fossils. But with a skeleton, a whole new anthropological world opens up.

I pondered on that world as the excavation progressed. I wondered what new knowledge the Turkana boy would bring us, knowledge that for the first time would allow us to understand the extinct *Homo erectus* almost as well as if we had met a living individual. How humanlike a childhood had the Turkana boy experienced, for instance? And how humanlike an adulthood had he been denied by his early death? All this knowledge lay in the future, or so I hoped. I wasn't to be disappointed.

The story of the discovery of *Homo erectus* is one of the epic tales of paleoanthropology, and it involves so many unlikely circumstances that it is amazing there is a tale to be told at all. Part of it is fact; some of it is fiction.

The story goes like this. A young Dutch anatomist, Eugène Dubois, developed a passion for finding the true human ancestor, the "missing link," in the parlance of his time, the 1880s. And he found it. The story goes on to relate how Dubois, having unearthed his paleontological treasure, was so dismayed by the negative reaction of the anthropological community that he turned his back on science, slipped into a kind of madness, and eventually came to believe that he had found not the missing link, but a giant gibbon instead.

47

At the time that Dubois first became obsessed with finding the missing link, few fossil humans were known. All the famous fossil discoveries in South and East Africa lay decades ahead. The only fossils of early human ancestors that had been discovered in the 1880s were those of Neanderthals, the first bones of which had come to light in a limestone quarry in the Neander Valley, near Düsseldorf, Germany. Neanderthals were a relatively modern form of human, which became extinct some thirty-four thousand years ago. Dubois was interested in what he thought was the earliest human form, something more primitive than Neanderthal man.

He had been inspired in his anthropological passions by the writings of the German zoologist Ernst Haeckel. Haeckel, a leading figure in European scientific circles toward the end of the nineteenth century, was a champion of Charles Darwin's ideas on evolution. His own books on evolution were widely read and influential. Like Darwin, Haeckel recognized that humans and apes have a common origin, linked by what he described as "Human Apes." He wrote, "The certain proof of their former existence is furnished by the comparative anatomy of Man-like Apes [the great apes] and Man."

Although humans and apes are closely related, there must have been some kind of intermediate link between them, concluded Haeckel. His reasoning was that the human capacity of speech surely required more than a single evolutionary step in which to develop. He called that intermediate form "Ape-like Man," or *Pithecanthropus*. The creature would have looked human in many ways and had humanlike mental characteristics, he suggested. However, it "did not possess the real and chief characteristic of man, namely, the articulate human language of words," Haeckel wrote in 1876. He therefore gave his Ape-like Man the species name *alalus*, which means silent: *Pithecanthropus alalus*.

Haeckel proposed that *Pithecanthropus* had arisen on Lemuria, a continent then thought to have sunk beneath the Indian

Ocean. From Lemuria, the evolutionary descendants of the creature would have migrated westward to Africa, northwestward to Europe and the Middle East, northward to Asia, and over the land bridge to the Americas, eastward via Java to Australasia and Polynesia. Bizarre though this global geography seems to us now, in Haeckel's time the basis of continental geology and plate tectonics was unknown, and the idea of extensive land bridges and vanishing continents was part of conventional scientific thought.

When Dubois completed his medical studies in 1884 his interest was to find *Pithecanthropus*, but he needed a way of getting to the lands where Ape-like Men once lived. Luckily, Holland's colonies included Indonesia, on the fringes of the supposed lost continent. Leaving behind a career in academic anatomy at the University of Leiden, Dubois took advantage of the presence of the Dutch East India Army in Sumatra, and secured a post as medical officer. He hoped this would give him the opportunity to organize the search for fossils. In 1889, two years after arriving in Java, he had convinced the military authorities of the value of his quest and was granted permission to devote himself to it. He was even given support to organize a paleontological survey, and this included gangs of convicts to do the excavating for him.

The survey initially covered sites in central Java, where thousands of animal fossils were recovered. Compared with today's moderate and careful affairs, Dubois's excavations were monumental and hazardous, often involving gangs of up to fifty of his "helpers." In August 1891 Dubois ordered that the survey be moved to the ancient sediments cut through by the Solo River, near the village of Trinil. During the three months of excavation that were possible before seasonal rains raised the river level above the site, two key fossils turned up in the cache: a tooth and a skull cap, both definitely primate.

The find convinced Dubois that he was on the right paleon-

49

tological track. He decided that the fossils belonged to a species of chimpanzee with some humanlike characteristics, and he named it *Anthropopithecus troglodytes*. But the next year's discoveries caused him to change his mind. Again employing gangs of convicts, Dubois shifted the excavation site upstream from the 1891 season. This time a femur, or thigh bone, was found. It looked virtually indistinguishable from that of a modern human. In fact, many anthropologists today think it is a modern human femur. Dubois believed that the tooth, skull cap, and femur came from the same individual. Still not certain that he had found Haeckel's missing link, Dubois named the creature *Anthropopithecus erectus*, a species of upright chimpanzee that, he said, "approaches closer to man than any other anthropoid."

After a year of worrying over the fossils, of being impressed by the large brain size and the upright stance implied by the thigh bone, Dubois changed his opinion. In 1893 he called the find *Pithecanthropus*, Haeckel's missing link. But because he couldn't be sure whether this early form of human had been able to speak, he didn't use Haeckel's species name, *alalus*. Instead, he called it *Pithecanthropus erectus*, because there was no doubt that this apelike man had walked erect. Later, the name was changed again, to *Homo erectus*, upright man.

Dubois said that his Java man was "the transition form which in accordance with the teachings of evolution must have existed between man and the anthropoids." Many of his colleagues were unconvinced. Although Haeckel agreed with Dubois that his missing link had been found, he considered the Java remains too fragmentary for anyone to say much about the extinct creature. Haeckel's position was a minority. Rudolf Virchow, the influential German scientist, doubted that all the fossils came from the same individual, an opinion accepted by many. Among those who did believe they came from one creature, many said it was more manlike than Dubois claimed; others considered it more apelike. Dubois lost on all counts.

It is interesting and instructive that a single set of fossils could provoke such contradictory expert opinion. Fossil anatomy can be extremely difficult to interpret, especially when it is fragmentary, as it so often is. People's expectations, their scientific preconceptions, influence their judgments. All scientists work from some kind of theoretical framework and interpret evidence in its light. Weak evidence can often be made to fit such a framework, whatever its form. I've seen that happen many times in paleoanthropology today.

Dubois faced this spectrum of interpretations wherever he went, and became intensely frustrated that he could not get other scholars to agree with him that *Pithecanthropus* was indeed a transitional form between ape and man. The scholars praised Dubois for the enterprising spirit and persistence that resulted in the discovery of such interesting—if nonhuman—fossils, but this praise was of little solace to Dubois.

There are two versions of what happened next in Dubois's life. The first, a familiar tale in the annals of paleoanthropology, is apocryphal. Angry and dismayed by his colleagues' intransigence, Dubois withdrew from the scientific community, and for more than two decades hid the fossils to keep other scholars from studying them. (There are several versions of where he stored them, including under the floorboards of his dining room and in a box in the attic. These differences are perhaps an indication that we are dealing with a popular myth.) Then, it is said, in frustration, anger, and a touch of madness, Dubois declared the fossils to be nothing more than those of an extinct giant gibbon.

A recent piece of research into Dubois's life, by Bert Theunissen, shows that this dramatic tale is baseless. Although Dubois was intensely irritated and upset that his Java man was not hailed by his peers as the missing link, and though he eschewed paleoanthropological debates after 1900, he did not withdraw from science. Nor did he change his mind about *Pithe-*

51

canthropus being a link in human evolution. The real story is more interesting than that, more scientific.

As an anatomist, Dubois's real interest was in brains, specifically the size of a species' brain in relation to the size of its body. These days the subject is a thriving issue in modern evolutionary and behavioral biology, but a century ago Dubois was a lonely pioneer. Theunissen has shown us that Dubois published far more on brain evolution than on *Pithecanthropus*, for which he is better known. The two subjects were not separate in Dubois's scientific life, however, because an understanding of brain size was crucial to determining the status of *Pithecanthropus*. "It was to obtain a better insight into this new organism that, soon after the discovery, I undertook the search for laws which regulate cerebral quality in Mammals, a study which indeed furnished evidence as to the place of *Pithecanthropus* in the zoological system," he wrote in 1935, six years before he died.

According to Dubois, with each major evolutionary advance, the brain doubles in size relative to the body, an idea he based on his understanding of embryology. The brains of great apes are a quarter the human size; the brains of carnivores and hoofed herbivores are an eighth; the brains of rabbits are a sixteenth, and so on. In this scheme, Dubois recognized that there was a gap: something should fit halfway between the great apes and humans, something with a brain half the size of the human brain. For *Pithecanthropus* to be this missing link it would have to slide in between the two, at precisely half the relative brain size of humans. The problem was, the brain of *Pithecanthropus* measured 855 cubic centimeters, that is, two-thirds the size of a modern human brain, not half. This would put the Java fossil creature outside the evolutionary progression toward *Homo sapiens*.

Dubois therefore indulged in a circuitous line of reasoning that, to him, rescued the status of *Pithecanthropus* as a human ancestor. If *Pithecanthropus* had a humanlike body, then its brain would indeed be too big for that of a direct human ancestor. But if its body had been much larger, say 220 pounds rather than the human average of 132 pounds, then its brain size relative to its body size would be reduced. In fact, it would then fit the required halfway position between great apes and humans, and would provide a jumping-off point for the evolution of true humans. Satisfied that he had achieved his aim, Dubois wrote in 1932, "I still believe, now more firmly than ever, that the *Pithecanthropus* of Trinil is the real 'missing link.'"

Because Dubois applied the name "giant gibbon" to this creature, many people took it to mean that he no longer considered

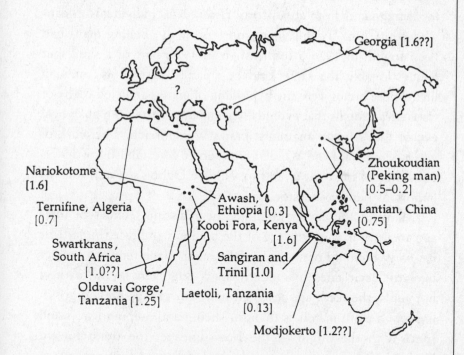

Homo erectus was the first human species to move out of Africa, certainly by one million years ago and perhaps significantly earlier. The map shows the major sites of discovery, with dates (in million years).

his *Pithecanthropus* to be linked to human ancestry. Not so, but the myth persists. In truth, there is no evolutionary law of doubling of brain size. Dubois had made a simple, but fundamental error in believing there was. Had he not made this error, he would have had no need to engage in his tortuous argument, and the idea of a giant gibbon would never have been born. But that is the nature of science and history.

Gradually, however, Dubois's claim for Java man gained more credibility, which encouraged another Dutch anatomist, Ralph von Koenigswald, to follow in his footsteps. In 1937 von Koenigswald went to Sangiran, another area of Java, and began excavations—interrupted by the war—that ultimately amassed fossil fragments from about forty *Homo erectus* individuals. Meanwhile, in China, Peking man (more properly, Beijing man) had been found: in 1926 a tooth, then in 1929 part of a skull, but unquestionably the same kind of creature as Dubois's missing link. These bones were the beginning of an unparalleled cache of *Homo erectus* fossils that eventually came under the keen analytical eye of the German anatomist Franz Weidenreich. The work of von Koenigswald and Weidenreich served to establish the ancestral status of *Homo erectus* and to validate Dubois's search for this important human ancestor.

Many of the earlier hominids bear distinct echoes of their ape ancestry in the structure of the cranium and face. In particular, early hominid brains are small and the face tends to project forward, as it does in modern apes. But even in the earliest hominids, the teeth are designed more for crunching and grinding tough plant materials than for slicing and pulping leaves and fruit. With the origin of the *Homo* lineage, the trend toward bigger grinding molars became reversed, not to fruit-processing teeth again, but to the teeth of omnivores, animals that may have included meat in their diet. This we see in *Homo erectus*.

The salient characteristic of this species, which allows us to recognize it from small fossil fragments, is the long, low cranium,

housing a brain two-thirds the size of modern brains, and sport-
ing prominent ridges of bone above the eye sockets, the so-
called brow ridges. The forehead is flat, and the back of the
cranium is curiously bun-shaped. Although the face juts out more
than it does in modern humans, it does so less than in earlier
hominids and in apes. When I hold a *Homo erectus* cranium in my
hand and look at it full face, I get a strong feeling of being in the
presence of something distinctly human. It is the first point in
human history at which a real humanness impresses itself so
forcefully.

It is true, I know, that the probable immediate ancestor of
Homo erectus, a species called *Homo habilis*, is in many ways simply
a smaller-brained version. And it is true that when I look at a
skull of this species, I cannot mistake it for that of an ape or one
of the small-brained early hominids. But in a sense difficult to
explain, *Homo erectus* seems to have "arrived," to be at the threshold of something extremely important in our history.

Even before the Turkana boy turned up at Nariokotome, the
great significance of *Homo erectus* in human history had been well
established, but with only a fraction of its anatomy known. The
hundred or so individuals known from various parts of the world
were represented for the most part by fragments of skulls and
jaws. Even the famous cave site of Beijing man yielded mainly
skull fragments. The other part of the skeleton most commonly
found, if any was found at all, was the thigh bone. This is because the femur is a robust bone that has the best chance of
surviving the vicissitudes of, first, burial and fossilization, and
then natural erosion out of the sediments, and recovery. Very
similar to the modern human femur, this bone in *Homo erectus*
nevertheless shows all the signs of belonging to a physically
active species: the bone itself is heavily buttressed and the sites
of muscle attachments are prominent.

Apart from the skull fragments, thigh bones, and various
other bones, such as parts of the pelvis, only a fraction of the

overall anatomy of *Homo erectus* had come to light—until the Turkana boy came along. The usual fate of an individual when it dies in the wild is that its carcass is scavenged. Hyenas, wild dogs, and even porcupines remove parts of the skeleton and leave their tooth marks on others. What remains of the skeleton dries out, is trodden on, kicked about, and variously dispersed by passing herds. Sometimes individual bones become buried, and if favorable chemical conditions prevail, they may become fossilized. Overall, the prospects of a single bone from a single individual entering the fossil record are small. The prospects of an entire skeleton entering the fossil record are modest in the extreme.

The key process in the entire business is burial: if bones become buried quickly after death, the chances of fossilization are greatly enhanced. And burial depends on there being a prevailing accumulation of sediments, such as occurs on the floodplains of rivers and lakes. The water carries fine sediments, which, under the most favorable conditions, can quickly entomb fresh bone. We anthropologists pray that one day an ancient Pompeii may be discovered, with a family of hominids entombed in volcanic ash where they sat at the moment of eruption. Failing that, we hope to find the remains of an individual who died near a river or lake edge. At Lake Turkana, and elsewhere in East Africa, such as the Olduvai Gorge and the Hadar region of Ethiopia, anthropologists are doubly blessed. Not only were there lake and river systems that give the best possible chances of fossilization of hominids and other creatures in past ecological communities; but these same habitats were also important in the evolution of our family. To the west of the Great Rift Valley, luxurious forests provided ideal habitats for apes of the quadrupedal variety. But for bipedal apes—the hominid family—the mosaic habitats created by the rift system were crucial, in their daily lives as well as in their evolutionary origin.

Our primary evolutionary adaptation, to bipedal locomotion, was a response to the need to forage for food in open environments, where patches of food were widely spread apart. At least as important, mosaic habitats—that tight mix of semidesert, savannah, woodland, riverine gallery forest, lowland, highland that is so characteristic of parts of the Great Rift Valley—increase the chances that new species will arise. The isolation of populations in separate geographical regions promotes the genetic revolutions that may give rise to new species and new adaptations.

During the three weeks in the fall of 1984 at Nariokotome, we made anthropological history: we slowly pieced together a skeleton of *Homo erectus*. Our excavation was progressing roughly from west to east, heading toward the lake and cutting deeper into the hillside as we went. The fossils themselves were on a level, ancient surface that our excavation was now exposing. This meant that for every yard we cut into the slope of the hill, we had to move more soil, rock, and debris above it to expose that ancient surface. In the end, we found ourselves clearing thirty square yards, shifting fifteen hundred tons of this so-called overburden. The excavation was not all toothpicks and camel-hair brushes. Gradually, as we exposed the bones and the ancient surface on which they rested, the circumstances of the boy's death became clear.

"The bottom of the bed was hard, a hard sandy bed in which reeds had grown," explains Alan. "There had been shallow water. We thought at one point it had been at the edge of the lake, but it was very low energy, not much water movement. There were no ripples in the ancient sand surface. You see ripples when there has been movement of water. Now we think it was either a lagoon near the edge of a lake, or maybe an oxbow of a river." Often when we find hominid fossils, we find remains of other animals too. Not this time. Just a few fish, some snails, and a piece of monitor lizard. But there were signs in the sand of other animals having walked by.

"After he died, the boy was lying face down in the shallows, head bobbing about in the water," continues Alan. "After a few days, or a week at most, the flesh was putrefying, and the straight-rooted teeth began to fall out. Hippos and other animals wandered by, and the skeleton, becoming disarticulated, got kicked about, the lighter bits going closer and closer to the shore. Something must have stood on the boy's right leg, because the fibula snapped in two, one of the pieces being pushed into the sand in a vertical position." The relatively light cranium floated or got kicked farther than anything, and then came to rest. A million and a half years later, a tree would grow in that exact spot: it was the wait-a-bit thorn tree that had caused us so much trouble early in the excavation.

About twenty years ago, a seed of the wait-a-bit thorn found its way into the sand-filled bowl of the boy's cranium, resting upside down a foot below the land surface. The boy was on his way into the world again, having been buried under tons of lake, river, and wind-blown sediment for a million and a half years. The immediate agent of his excavation was the little stream that cut a gully through the ancient sediments. The deeper the stream cut, the more the sloping land surface eroded away. Now that the boy's cranium was near the surface again, moisture could reach it. And the upside-down cranium served to trap moisture, providing a favorable spot for the seed to germinate. As the tree grew, its roots penetrated into the soil, and the cranium shattered in slow motion, the fragments held in the ancient sediment.

I don't like cutting down trees when we are in the field, on principle. It's the same principle that makes us careful to restrict our Land Rover tracks. The environment here is so fragile that even minor disturbances can lead to rapid erosion. However, when we discovered that our prized cranium was entangled in the tree roots, the press of paleontological necessity overcame the immediate environmental principle, and the tree came down.

Why did the boy die? We don't know. The situation of his burial was the inspiration for the story we told earlier. But, apart from the evident fact that a *Homo erectus* boy did indeed die in or near a lagoon, the rest of the story was pure imagination, based on elements of what we think we know about the life of *Homo erectus*.

We do know that there are no marks of carnivore gnawing on his bones, so he was probably not living prey to a carnivore, nor was his carcass scavenged as he lay in the water. "The only sign of disease is a tiny bit of resorption of the gum, on the mandible, where the deciduous second molar had been shed," says Alan. "Often infection sets in when the gum is broken with the new tooth. It's possible that he got sepsis and died." It may sound unlikely, but, as Alan found when he was looking at some sixteenth-century parish records for St. Martin's-in-the-Fields, in London, life and death before the age of antibiotics were very different from what we know. The leading cause of death was the plague, perhaps to be expected. But septicemia as a result of tooth infections came second. Interesting, but inconclusive.

Midway through the excavation I had to fly to Nairobi for a few days on museum business, an irritating diversion I'd rather have avoided but couldn't. While I was there I sent a telegram to Pat Shipman, Alan's wife, who was back in Baltimore: "Be advised that we are in the process of excavating an *erectus* skeleton. It's fantastic, and Walks [a nickname I sometimes use for Alan] wanted you to be among the first to know." Little did I know that one of the visitors to the camp was already telling the Nairobi *Time* magazine office what we were finding. Our secret was to be out sooner than we wanted.

On September 18 I returned to Nariokotome, and Alan could barely wait to show me what he had. Even from fifty yards, as I approached the excavation, I could see why he was excited. Newly exposed leg bones lay on the surface, and they made a pretty sight. They were the bones of the right leg: the thigh bone and the tibia of the lower leg. Immediately we could see that what we had begun to suspect about the boy was indeed true. He was very tall. "It looks as if some people are going to have to eat their words," commented Alan. "No doubt about it."

One of the frustrations of dealing with fossil fragments is that we can never be sure what the whole animal looked like. That's one reason the Turkana boy is such a spectacular find. Before he came along, anthropologists had tried to estimate the height of *Homo erectus* by using single bones, maybe a thigh bone from one site, an arm bone from another site—that kind of thing. But a true estimate of stature can be obtained only from the leg bones of a single individual, the thigh bone and the shin bone. Measured against modern human bones, these fossils can give a good indication of the height of a *Homo erectus* individual.

Estimates of *Homo erectus* stature made from isolated bones found before the Turkana boy depicted a stockily built, heavily muscled, not very tall creature. For instance: "They were exceptionally powerful people, both the men and the women, with muscles to match their thick frames," wrote Donald Johanson, in his book *Lucy.* "Although a *Homo erectus* male would have been too small to excel as a professional football player, a properly motivated one probably would have been devastating at lacrosse or hockey, two of the most physical sports now played by medium-sized men."

We could see that the Turkana boy was anything but medium-sized, given his young age: his leg bones were clearly those of a tall individual. We began to estimate his height, measuring the different segments of his body and making allowances for the fact that the soft ends of the bones, the unossified epiphyses, had been lost in the fossilization process. We came up with a figure of between five feet, four inches and five feet eight. "He was a strapping youth," said Alan, "and he would have grown up to be well over six feet." A strapping youth. I liked the phrase, and I thought about what I would say in our press announce-

ment.

Perhaps he was an exception, a freakishly tall boy? It is possible, but the odds are against it. The individuals most likely to come to light in a fossil find are the most common ones, those in the middle of the statistical distribution. To buttress our case, we can refer to fossil 1808.

The number 1808 is the museum accession code of a bizarre fossil we had found a decade earlier, on the east side of Lake Turkana. Described simply as "associated skeletal and cranial elements" in one standard hominid fossil book, 1808 gives us a glimpse into the last weeks of life of a *Homo erectus* individual, possibly a female, who lived about the same time as the Turkana boy, but on the other side of the lake. It is reasonable to guess that a couple of weeks before she died, 1808 ate part of the liver of a large carnivore, a lion or a hyena, perhaps. The signs are in her long bones, which are shattered, as are the rest of her fossilized remains.

The clue is that the surfaces of the long bones are not smooth, as would be expected. They are covered by a thin, rough layer, the result of a brief period of bleeding from the surface of the bone, followed by rapid bone formation, and then death. Alan once took a thin section through one of these bones and showed it as a slide to a group of clinicians at The Johns Hopkins Medical School in Baltimore. Their diagnosis was unequivocal: hypervitaminosis A, or the result of ingesting too much vitamin A. The clinicians were astonished when they learned that the bone they were looking at was a million and a half years old; they said it was indistinguishable from clinical cases they see today. Hypervitaminosis A can develop if an individual eats huge quantities of vegetables rich in the vitamin, such as carrots, or, what is more likely, carnivore livers, the richest known source of vitamin A.

As well as having a fatal desire to eat carnivore liver, 1808 was tall, about five feet nine, as far as we can tell from her shattered remains. It is surely no coincidence that the two *Homo erectus* individuals whose height we can estimate were both tall.

61

This gives us confidence to say that, unexpectedly, we were dealing with an exceptionally tall species.

We were even more surprised by our colleagues' reaction when, later in the year, we announced the discovery of the boy, and said, "He was a strapping youth, and that's a surprise . . . *Homo erectus* clearly was taller than we had imagined." "But we knew that all along," came the chorus, including Don Johanson. "We've had pieces from China and Java that suggested that these individuals approached six feet in size," Johanson told the *New York Times*. "I don't think this aspect of the discovery is so astonishing." When he read this, Alan laughed, pulled out a copy of *Lucy*, and sent a copy of the lacrosse quote to the *Times* reporter who wrote the article.

The reaction to our statement about the boy's size means, I believe, that people's ideas were generally so unformed that it was easy to believe what seemed to be true at the time. As more of the Turkana boy's skeleton emerged from the ground, all doubt about the stature of *Homo erectus* dissipated. The fossils from Java and China that Don referred to had been discovered many years before, so he wasn't pointing to new data when he said *Homo erectus* was tall. He was looking at old data, but with a new perspective, one that clearly was influenced by the Turkana boy's skeleton.

With the end of the digging season fast approaching (dictated as much by our money running out as by anything else), we were forced to assess how we should proceed. "Richard and I stared at the site plan over lunch and decided that if the rest of the skeleton is there, it is dispersed over a wide area," recorded Alan, on the evening of September 19, three days before we pulled out. "Consequently, the whole of the hillside all round will have to be leveled."

This was not a pleasing prospect. Although we would dearly have liked to get the missing pieces—some arm bones, a few teeth, but particularly the bones of the hands and feet—we knew we might spend a lot of time and money and not find anything. We feared that the hands and feet, light bones that they are, might have been trampled and kicked to where the ancient bank was. This was near to where the thorn tree had been, and right where the land surface was being eroded away by the action of the little stream. "Lost forever as sand into the Nariokotome" was how Alan described the probable fate of these bones.

"Everyone is a little saddened," I wrote in my diary the day before we packed up camp. "We all had high expectations, but a fully complete skeleton has eluded us." Two weeks later in Nairobi, our sadness evaporated as the individual bones were cleaned and the skeleton began to be assembled. "Come and look," Alan called one day. "Come and see your Turkana boy." Alan and Emma Mbua, curator of the hominid fossils at the museum, had managed to put the boy's skeleton in a standing position. It was to be photographed against a dark background for an article Alan and I had written for *National Geographic*. "Remarkable" was all I could say at first. "Remarkable."

There, standing erect and tall before me, was a virtually complete human ancestor who had lived more than a million and a half years ago. How very human he looked in that setting! It was a moving moment for me, different from the keen thrill of the excavation just weeks earlier. This was a deeper emotion, one derived from the broad sweep of human prehistory that I am privileged to see in my work. I realized I was face to face with a link in the chain that joins me, today, with the earliest human ancestors, apelike creatures who lived perhaps 7.5 million years ago. I could sense that, in our search for emerging humanness during that huge swath of evolutionary time, the Turkana boy has some of the answers we seek. The excavation of his bones was paleontological bliss. Having him yield those answers to us was going to provide a way of touching the very substance of our history.

The serious moment passed, and photographs of the boy

were taken, including one in which Emma stood by his side, the shorter of the two humans in the picture. Yes, the Turkana boy is a strapping lad.

Less than a month after we closed the season at Nariokotome we were poised to announce publicly our discovery. Normally, scientists like to have completed at least a preliminary analysis of a new find and to have presented it to their colleagues. This time, we were forced to rush the public announcement because the *Time* magazine office in Nairobi, tipped off about our find, could wait no longer before going to press with the story. Alan and I arranged to have simultaneous press conferences on October 18, mine in Nairobi at the museum and his in Washington, at the National Geographic Society.

As I was preparing for the press announcement I thought once again of the image of the boy as he had stood before me two weeks earlier. Anthropologists mostly think about fossils in terms of detailed anatomy, of branches on an uncertain family tree. Only infrequently do they regard the fossils as animals, individuals that long ago faced the daily challenge of surviving and thriving. This is understandable, because typically the fossils we find are so fragmentary that one can do little more than that. The impression of the Turkana boy as a person, however, is so intense that it is impossible not to think of what his life had been like. How tall would he have been as an adult? Did he spend his childhood with his siblings, learning the exigencies of the *Homo erectus* life? What would that life have entailed? What degree of language ability did he have? Did he have a sense of self, of introspective awareness, as you and I do today? How *human* was he?

With these kinds of questions turning over in my mind, I sat down in front of the television camera, prepared to describe to the world one of the most remarkable hominid fossil discoveries of all time.

PART TWO

In

Search

of

Beginnings

4

Of Myths and
Molecules

H o m o   e r e c t u s, the Turkana boy's species, represented a pivotal point in human evolution. More or less everything that preceded *Homo erectus* was distinctly apelike in important respects: in some of the anatomy, life history, and behavior. And everything that followed *erectus* was distinctly humanlike. The Turkana boy had been part of a major shift in human evolution, one in which the seeds of the humanness we feel within us today were firmly planted. In addition to important changes in overall body form and life patterns, *Homo erectus* was at the forefront of a surge in brain size, a boost in mental capacity. It was, I believe, at the real beginning of the burgeoning of compassion, morality, and conscious awareness that today we cherish as marks of humanity.

We need to view this pivotal point in temporal and biological perspective, taking into account the time at which the human family first appeared and the debates surrounding the date of that beginning.

The "story" of human origins goes something like this. Once upon a time, a very long time ago, a species of unusual ape in Africa was forced out of its traditional forest home because a cooling climate had steadily reduced the forest cover. Our resourceful ape grasped the ecological opportunity and, in its new, open-country niche, at once began to undergo a series of evolutionary changes. Gradually it came to stand and move on two legs, not four; to make and use stone tools and weapons; to reduce the size of its daggerlike canine teeth; to enlarge the size of its brain.

A positive feedback system was set up, each development leading to the next: the more it stood upright, the more it could use its hands; the more it used its hands, the more it needed to be upright; the more intelligent it became, the more it could rely on stone-tool technology. Eventually it became a primitive version of us, erect and intelligent, a skilled tool maker, an accomplished hunter. It stood triumphant on the plains of Africa, leaving less resourceful apes to skulk in the receding forests.

This was the dominant fantasy—and I use the word advisedly—that held sway in anthropology for a long time, not least because it seemed to be a plausible account. The major elements of the account are two: first, the attainment of humanity required initiative and effort, and the apes remained apes because they didn't exert themselves enough. Second, the evolutionary transformation from ape to human was effectively instantaneous, because the three qualities we take as separating us from the apes— upright walking, tool making, and high intelligence—all began to appear right from the beginning. In other words, the first member of our family was already identifiably like us, albeit in primitive form. These two elements, I believe, tell us something about ourselves, something about what the quest for human origins means to professionals and nonprofessionals alike.

The first element of the fantasy—the idea of initiative and effort in human evolution—was explicit in anthropological writings in the early decades of this century, but happily is no longer apparent. Even though anthropologists nowadays no longer think of initiative and effort as being part of the evolutionary process, there is still a reluctance to view the whole process as ruled by chance and circumstance. The demise of the second element—that the very first hominid was already human in important respects—is much more recent. Evidence from the prehistoric record finally forced a recognition that being a hominid does not automatically equate to being human, however primitive.

A 1922 essay by Edward Grant Conklin, a professor of biology at Princeton University, gives clearly and concisely the sense of what evolution was considered to be about at that time: "The lesson of past evolution teaches that there can be no progress of any kind without struggle." Grafton Elliot Smith, a leading British anthropologist of the same era, wrote that our ancestors "were impelled to issue forth from their forests, and seek new sources of food and new surroundings on hill and plain, where they could obtain the sustenance they needed." Elliot Smith also characterized human evolution as "Man's ceaseless struggle to achieve his destiny," leaving no doubt that becoming human was a prize to be earned; and, incidentally, that there was a prized goal—his destiny—to be reached.

So much for humans, but what of our close cousins the apes? Anthropologists of this early era were clear about the difference between us. "Why, then, has evolutionary fate treated ape and man so differently?" asked Arthur Keith, a contemporary of Elliot Smith's. "The one has been left in the obscurity of its native jungle, while the other has been given a glorious exodus leading to the domination of earth, sea, and sky." Elliot Smith's explanation was as unequivocal and scalding as a bad year-end school report: "While Man was evolved amid the strife with adverse conditions, the ancestors of the Gorilla and Chimpanzee gave up the struggle for mental supremacy because they were satisfied with their circumstances." In other words, the apes had the same evolutionary opportunity that we had, but blew it through indolence. No pain, no gain.

While these sentiments may sound ludicrous to our modern ears, we have to remember that they were the serious commentary of the most prominent scientists of the time. Their position rested on two sets of assumptions, one scientific, the other social. The scientific were formed by looking at an incomplete fossil record. We now know that the evolutionary history of most groups—such as the hominids or other large mammals—can best be described as shaped like a bush, with many branches terminating in dead ends. These dead ends are the species that go extinct at different times.

The important point here is that the probability of a particular species going extinct is determined as much by external factors, such as a catastrophic change in habitat, as by internal factors, such as how well adapted or fit they are. Equally, the chance that a particular species will diverge evolutionarily, producing two daughter species, is as much determined by external circumstances as by properties of the parent species.

The upshot of all this is that the species that survive and change through time are as much influenced by good luck as by good genes. But if a paleontologist is able to sample, say, only 10 percent of the species that actually existed within a group, he or she will come up with an incomplete picture of the evolutionary bush, one in which it may look as though there were steady trends through time, trends like the increase in the size of body, the length of the tooth row, and the size of antlers. The classic example is the evolutionary history of the horse, for so long described as being the result of steady trends of increase in body size, reduction of the number of toes, change in the structure of teeth, and so on. These days, with a more complete fossil record, we can see that the evolutionary history of the horse was the classic bush, a series of stochastic speciations and extinctions, not an inexorable trend. Often, for example, the body size decreased in some of the branches, and these branches just happened to go extinct.

Trends, real or imagined, have a powerful and seductive sense of inevitability about them. They give the impression that the anatomical feature in question is being driven in a particular direction. Add this to the social environment in which Elliot Smith and his colleagues were commenting on human evolution, and a very particular perspective comes through. The social context, at the end of the last century and the beginning of this century, the high point of Victorian and Edwardian expansionism, was indeed success through effort. There were great rewards to be had in the newly emerging industrial age, but not for the indolent, only for those who exerted themselves. The same social ethic insinuated itself into scientific thinking, especially into views of evolution.

William King Gregory, a major figure in anthropology during the early decades of the century, took a skeptical look at the writings of Elliot Smith and others regarding man's supposedly elevated place in nature. "The erect posture which has enabled man to look down on the world of quadrupeds may well be one of the bases of man's colossal and impregnable superiority complex," he wrote in 1928. "The contrary notion, that man is an uptilted and still only partly refashioned four-footed animal, is to this day deemed impious and even blasphemous by many accredited spokesman of millions of people in Boston, Dayton, and points west and south."

Gregory, of the American Museum of Natural History, was one of the more far-seeing thinkers of his time and did constant battle with his boss at the museum, Henry Fairfield Osborn, who could not bring himself to accept that apes had anything at all to do with human evolution. Gregory's comments about the source of our superiority complex—our standing on our hind legs—may have some merit. More likely, what was at work here was the intrusion into scientific theories of the Victorian-Edwardian work ethic. Gradually, however, this particular social overlay on scientific explanation was removed. But even when the more florid phraseology of Elliot Smith and his contemporaries disappeared from anthropological texts, something of the substance remained, like the Cheshire cat's smile.

The apes were no longer described as failures through their own lack of effort, but were implicitly considered to have come to a full stop in terms of evolution. David Pilbeam, who is now at Harvard University, explained it this way: "It was generally assumed that the great apes were closely related to each other, somehow primitive, and that humans have done all the evolving since the last common ancestor, while the apes have changed hardly at all." In other words, look at a chimpanzee, and you are looking at a living version of our distant ancestor. This is an incorrect assumption.

The first thing we learn when we work with fossils long enough is that species change through time, sometimes subtly, sometimes dramatically. And the second thing is that the creatures of the past are not simply primitive versions of extant species. Alan Walker has recently been working with a fossil species that beautifully illustrates this paleobiological principle. The fossil is that of *Proconsul*, an apelike creature that lived in Africa eighteen million years ago. My mother had found the original specimen in 1948, an exquisite skull, on Rusinga Island in Lake Victoria. Since then many parts of the skeleton, and many different skeletons, have been recovered, and this is where the interesting story lies.

"*Proconsul* may well be ancestral to the later apes in Africa," says Alan, "but it is not an ape in the sense that we know them today. For example, some of the ankle bones are slender and monkeylike, but the big toe is robust and apelike. The same hybrid pattern is seen in the pelvis: the ilium, or upper portion, is like that of Old World monkeys, whereas the acetabulum [the

*F*lakes and choppers, typical of the Oldowan, made between 2.5 million and 1 million years ago. (Nicholas Toth)

*H*and axes and a cleaver (right), which characterize Acheulean assemblages, made between 1.6 million and 200,000 years ago. (Nicholas Toth)

*T*he Mousterian assemblage was based on large flakes, made between 200,000 and 33,000 years ago. (Nicholas Toth)

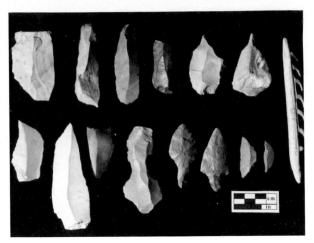

Upper Paleolithic assemblages contained a wide range of implements, many based on stone blades, some on bone and antler, like the harpoon shown here. (Nicholas Toth)

Donald Johanson displays the partial skeleton Lucy soon after its discovery, in 1974. (Cleveland Museum of Natural History)

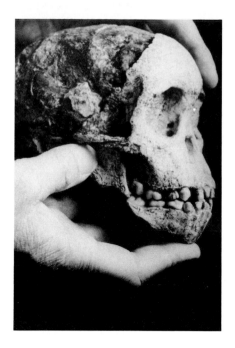

The Taung child skull, found in 1924, was the first early human ancestor discovered in Africa. (The University of the Witwatersrand)

A Cro-Magnon skull, representing modern humans in Europe; this one was found in the Dordogne region of France. (Milford Wolpoff)

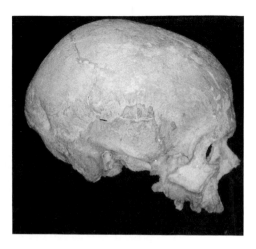

Skull 406, or Australopithecus boisei, *found at Koobi Fora, Kenya, has robust jaws and dentition. The species lived between* 2.6 *million and about* 1 *million years ago in East Africa.* (P. Kain/Sherma)

Skull 1813, *found at Koobi Fora, probably represents a branch of the human family tree, of about* 2 *million years ago, that is yet to be named.* (P. Kain/Sherma)

Skull 1470, a member of Homo habilis, found at Koobi
Fora, represents the first large-brained hominid species
known. This skull is just under 2 million years old, but the
species may have arisen as early as 2.5 million years ago
and continued to perhaps 1.6 million years ago. (P. Kain/
Sherma)

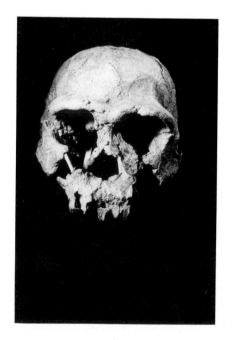

Skull 3733, a member of Homo erectus, found at
Koobi Fora. This species originated about 1.7 million years
ago and existed until about 300,000 years ago. It was the
first human species to move out of Africa. (P. Kain/Sherma)

This skull, often known as Mrs. Ples, represents the species Australopithecus africanus, and comes from the cave of Sterkfontein, South Africa. (P. Kain/Sherma)

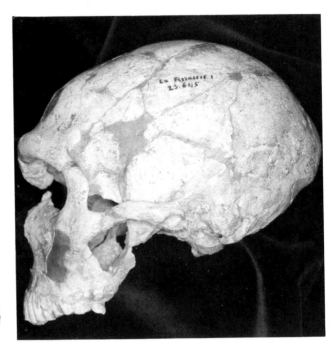

A Neanderthal skull from La Ferrassie, France. Notice the large brain case and protruding face. (Milford Wolpoff)

The part human–part animal figure in the Hall of Bulls in Lascaux. Squint, and the head sometimes looks like that of a bearded man. (Doc. N. Aujoulat, Dept. Art Parietal-CNP)

The mysterious scene in the Shaft at Lascaux. (Doc. N. Aujoulat, Dept. Art Parietal-CNP)

An example of San art, a supine therianthrope with fish. A small antelope, bleeding from the nose and therefore dying, stands on a double line of white dots. The image is part of a complex panel of illustrations. The lines fringed with white dots lead to other depictions (not seen here). (T. A. Dowson, with permission of the Director, South African Museum, Cape Town)

site of articulation with the head of the femur] is large and shallow, like that of the great apes. The wrist is similar to Old World monkey wrists, while the shoulder and elbow are remarkably apelike." In other words, this animal was a mosaic of features we now find in two entirely separate groups—two different Superfamilies, in biological parlance—in addition to some novel anatomy of its own. The obvious implication, as Alan notes, is that "just as *Proconsul* was not in the precise anatomical mold of a modern ape, neither was it behaviorally."

The lesson, here, is that if we wish to speculate on what the immediate ancestor of humans and the modern great apes was like, we cannot expect to find a precise living model, a living fossil. Be guided by modern anatomy, but don't be limited by it. Charles Darwin, and his friend and champion Thomas Henry Huxley, recognized anatomical links between humans and the African apes, the chimpanzee and gorilla. The African apes are our closest relatives, they said, so that is where we should look for guidance. But even so all is not what it may seem at first sight. There are subtleties in the formation of anatomical structures that we are only just beginning to understand. Unless we have a complete understanding of these subtleties, there is always the possibility of our making mistakes in inferring close evolutionary relations between species that share an anatomical structure. And it is inescapably true, of course, that paleontologists are limited to studying anatomical structures in the "hard parts" of species, the skeletons. We know that skin and hair, particularly in their color and pattern, are often different between species whose skeletal structure is similar, even identical. This is a piece of knowledge that we, who deal with animals only at the level of fossilized bones, simply have to live with.

Having identified a link between African apes and humans, Darwin went on to hypothesize that the human family arose in Africa. He was right. All the earliest hominid species have been found in Africa, and Africa alone. Only with *Homo erectus* did our ancestors expand beyond the African continent. This is yet one more reason for viewing *Homo erectus* as marking an important point in human evolutionary history.

Darwin also formulated the notion that a complex of human-like characteristics—bipedal walking, tool making, and an enlarged brain—evolved in concert. This is the second of the two major elements in the last generation's anthropological fantasy, that right from the beginning, hominids were already essentially humanlike in important respects. "If it be an advantage to man to have his hands and arms free and to stand firmly on his feet, of which there can be no doubt from his pre-eminent success in the battle for life, then I can see no reason why it should not have been more advantageous to the progenitors of man to have become more and more erect or bipedal," Darwin wrote in 1871, in his *Descent of Man*. "The hands and arms could hardly have become perfect enough to have manufactured weapons, or to have hurled stones and spears with true aim, as long as they were habitually used for supporting the whole weight of the body . . . or so long as they were especially fitted for climbing trees."

Our small, bipedal, weapon-wielding, savannah-hunting ancestor was in a position to develop greater intelligence through more intense social interaction, said Darwin. And the large canine teeth would disappear too. "The early male forefathers of man were . . . probably furnished with great canine teeth; but as they gradually acquired the habit of using stones, clubs, or other weapons, for fighting with their enemies, or rivals, they would use their jaws and teeth less and less. In this case the jaws, together with the teeth, would become reduced in size."

In the almost complete absence of fossil evidence, Darwin had erected a plausible outline, emphasizing the key characteristics of humanity as the prime movers in the evolutionary transition from ape to human. "For Darwin, the first evolutionary step our ancestors took away from the last common ancestor with the apes encompassed everything that later came to be identified—

and valued—as 'human,' " says David Pilbeam. "So plausible was it, so powerful an image, that it persisted until relatively recent times."

The persistence of this powerful evolutionary package played an important role in a dispute, now famous in the annals of human origins research, between anthropologists and biochemists over the origin of the human family. David was very much involved in this dispute, first as one of the chief anthropological protagonists, then as a champion of the biochemists. His shift in position had a major effect on our discipline by legitimizing what may be called molecular anthropology.

Ever since our science was established, we anthropologists have relied on fossil evidence to reconstruct human evolutionary history. We know that fossil evidence is not always easy to interpret; problems of interpretation are manifest in our changing theories about human history and our differences of opinion about particular fossils. Nevertheless, fossils were our most direct link with our past. Then, during the 1960s, another line of evidence was introduced: molecular data from genes and proteins in living creatures, such as ourselves and our closest relatives, the African apes.

The notion of using molecular evidence to probe issues of genetic relatedness is basically straightforward. Once a common ancestor has diverged into two daughter species, the genetic material in them will gradually accumulate mistakes, or mutations, and the species will become increasingly different from each other. The more distant in time the evolutionary divergence, the greater will be the accumulated genetic difference. And if the mutations accumulate regularly through time, there is what biochemists call a molecular clock: the changes through time caused by the accumulation of mutations. By measuring the degree of accumulated genetic difference between two related species, the time since they diverged from each other can be calculated.

In the early 1960s Morris Goodman, at Wayne State University, introduced this kind of molecular evidence into anthropology when he demonstrated the close genetic relationship between humans and African apes, the chimpanzee and gorilla, and the distance between humans and the Asian great ape, the orangutan. But it was the Berkeley biochemists Allan Wilson and Vincent Sarich who really caught the anthropological community's attention by suggesting, in 1967, that molecular evidence showed humans and apes to have diverged from each other some five million years ago. At the time, anthropologists believed that this divergence took place much earlier, at least fifteen and possibly as much as thirty million years ago.

Sarich at one point said provocatively, "One no longer has the option of considering a fossil specimen older than about eight million years a hominid *no matter what it looks like.*" His logic was as simple as it was challenging: any resemblance to hominid anatomical features in a fossil older than five million years (plus a couple of million years as a safety margin) must be purely accidental. Such a fossil might look like that of a hominid, but it could not be, he asserted, because it was too old.

No scientist likes to be told that his or her chosen professional approach is useless, least of all by someone outside that profession. It is no surprise, therefore, that Sarich's exhortation was received less than enthusiastically by most anthropologists. For a decade and a half, open hostility raged between the camps, with anthropologists vigorously attacking Sarich and Wilson's work and refusing to incorporate any of it into models of human origins. "All those years the paleoanthropologists functioned as if we didn't exist," Wilson laments. "Most people ignored us," adds Sarich, "or vilified our results and our method."

The first time I heard Vince Sarich talk was in 1983, at a conference on diet and human evolution, held in Oxnard, some sixty miles north of Los Angeles. Everything I had heard about him seemed to be true. He is an enormous man—in stature, voice, and ego. And he has a knack of irritating paleontologists like me. "The key to the past is the present," he boomed, which surely was a direct challenge to the utility of fossils, my stock-in-trade. "How can we expect to understand the past without studying fossils?" I wanted to know. Sarich replied with a version of one of his well-known aphorisms: "I know my molecules had ancestors, but you paleontologists can only hope your fossils had descendants." Here in brief was the debate in which Sarich and Wilson and the anthropological community had been embroiled for almost two decades.

The fundamental question over which anthropologists and biochemists debated so bitterly—when did the first member of the human family appear?—seems simple enough. But the practical issue for anthropologists was how to recognize such a fossil when it was uncovered. "We recognized it—or we thought we did—when we saw part of the Darwinian package," explains David. "From fragmentary fossil evidence we built a complete picture of the first hominid. We said it was probably bipedal, probably made tools, was social, and probably hunted too." The fossil to which all this was attributed was a fragment of upper jaw of a primate. The key feature was that the canine teeth were small, distinctly unlike those of an ape, but similar to those of humans. There were other features, too, that seemed to link the fossil with human ancestry, such as the shape of the cheek teeth, the thickness of the enamel on the teeth, and the supposed shape of the jaw itself.

"All you needed in a fossil was small canines and the belief that it was a hominid, and everything else followed as a functional package," David now explains. "This whole view reflects the expectation that the earliest hominid is already a pretty special creature, that it is already well on the way to being a human. It is very much a cultural animal."

This little fossil jaw, named *Ramapithecus*, had been found in 1932 by a young researcher, G. Edward Lewis, in fifteen-million-year-old deposits in India. But it wasn't until 1961 that *Ramapithecus* sprang to prominence as the putative first member of the human family. Elwyn Simons, then at Yale University, studied Lewis's fossil find again, concluded that *Ramapithecus* was a hominid, and published a landmark paper saying so. David soon joined Elwyn at Yale, and for more than a decade their names were inextricably linked to the promotion of *Ramapithecus* as the first member of the human family, which, they said, evolved at least fifteen million years ago, maybe even thirty million. The Simons-Pilbeam thesis quickly became the received wisdom among the profession, and, like most of my colleagues, I supported it. Then Vince Sarich and Allan Wilson entered the picture, and declared that Simons and Pilbeam and the rest of us anthropologists had got it badly wrong.

The way things worked out is now well known, of course. We anthropologists were forced to admit that we had been wrong and that Sarich and Wilson were closer to the right track than any of us had even imagined. Genetic measures made over the years following Wilson and Sarich's 1967 landmark publication, some using proteins, others using various forms of nucleic acid, all point to a recent divergence date, somewhere close to five million years ago, perhaps a little earlier. These days the date is often said to be "somewhere between five million and ten million years ago," with 7.5 million years a good average. In 1980 and 1982 important fossil ape discoveries reported from Turkey and Pakistan were consistent with what Sarich and Wilson had been urging on us for thirteen years. I remember saying at a lecture at the Royal Institution in London, "I am staggered to believe that as little as a year ago I made the statements that I made about the molecular clock data." I hadn't been as outspoken as some of my colleagues, but I had certainly been negative. "I think the molecular people are closer to the truth than we've ever given them credit for," I added, still a little conservative.

For David Pilbeam the experience was both difficult and salutary. "I am less sanguine than I used to be about the extent to which fossils can inform us about the sequence and timing of branching in [human-ape] evolution," he wrote in 1983. "You are much better off using molecular evidence if you want to be sure about the location and timing of branching points. And that is a difficult admission to have to make, for someone who was brought up to believe that everything we needed to know about evolution could be got from fossils. Fossil evidence is important, of course, but it can address only some of the questions we face."

What led him astray over *Ramapithecus*, says David, was similarity of anatomy. "We saw a few anatomical features that seemed to imply relationship, and accepted them uncritically." David and Elwyn were ensnared in a trap that lurks for all in our profession: similar anatomy does not always imply close evolutionary relationship. In evolution, identical anatomy may appear in two unrelated groups when they adapt to identical pressures of natural selection. The *Ramapithecus* episode served to make many of us a great deal more alert to the trap. Unlike David, however, I still have a lot of confidence in our ability to reconstruct evolutionary events based on fossils alone. Perhaps this reflects the stubbornness of someone to whom anatomy is much more familiar than molecular biology. But, difficult though it may be, I am sure that we shall be able to identify the invisible genetic relationships as they are embedded in the anatomy we see.

Vincent Sarich believes the episode tells us as much about the nature of our quest—understanding the place of humans in the universe of things—as it does about scientific methodology. "As I see it, the basic problem has nothing to do with the evidence, be it molecular or paleontological, but with the difficulty most of us have with accepting the reality of our own evolution," Sarich says. "We have developed sufficient intellectual maturity to make overt denial of the fact of human evolution impossible. Its positive acceptance, however, is made easier in direct proportion to the distance in time which separates us from our proposed ancestors."

I suspect there is some merit to the argument. Even with the rationality that is part of being human, we find it difficult—emotionally, at any rate—effectively to attach ourselves through an unbroken chain of genetic inheritance to the simian world. In the strong anti-evolution atmosphere that prevails in much of the United States, it would be surprising if professional anthropologists were not affected in some way, at least in unconsciously attempting to make the fact of our evolution more palatable by making the chain as long as possible, distancing humans far from the rest of nature.

The upshot of the past two and a half decades of biological and molecular anthropology is that we have a major landmark in human evolution: the origin of the hominid family, close to 7.5 million years ago, perhaps a little more, perhaps a little less. By the time the Turkana boy came along 1.6 million years ago, the human family had been around a long while.

5

Upright Apes and
Family Relations

F o r   m e , the fundamental distinction between us and our
closest relatives is not our language, not our culture, not our
technology. It is that we stand upright, with our lower limbs for
support and locomotion and our upper limbs free from those
functions. In essence, humans are bipedal apes who happened to
develop all these other qualities we usually associate with being
human. And if we think again about the molecular data, which
ally us closely with the chimpanzee and gorilla, then, no question about it, we are apes of a kind, "rather odd African apes," as
David Pilbeam once put it.

The prehistoric record in Africa is now extensive, no longer
the quip about fewer fossils than would cover a dining room
table. By my count there are fossilized fragments of about a
thousand human individuals from the early part of our evolution,
and I wouldn't even try to count the number of stone tools. All
this shows clearly that the earliest stone tools appear in the
record about 2.5 million years ago, some five million years after

the origin of the human family. Of one thing we can therefore be certain: the Darwinian package of bipedalism, tool making, and intelligence marching in evolutionary concert is not correct.

So important to our later evolutionary history was the freeing of our hands that my preference is to use the term "human" for the first bipedal apes. I know that many people become exercised about the implications of names, particularly when it comes to us, but for me, "human" and "bipedal ape" are synonymous. I'm not saying that once the bipedal ape had evolved, you and I were evolutionary inevitabilities, because evolution doesn't work like that. Nor am I suggesting that the earliest bipedal apes had the same intellectual powers or outlook as we do. Of course they didn't. All I am suggesting is that the origin of a bipedal form of locomotion was so fundamental a change, so replete with profound evolutionary potential, that we should recognize the roots of our humanity where they really are. I would make a distinction, however, between calling the first bipedal apes human and expecting to find humanlike behavior in these creatures. Once we recognize the importance in our history of the origin of bipedalism, we can begin to identify the emergence through evolutionary time of that intangible, indefinable, and yet intensely felt sense in us that we identify as true humanity.

We should not expect to get a lot of help in this respect from the prehistoric record, but we can look to it to learn something about the origin of bipedalism. If we are guided by the molecular evidence, then we know roughly when it came about: roughly 7.5 million years ago. What really interests us, of course, is why it evolved at all. What were the circumstances that favored its appearance? And what were its immediate consequences?

The earliest known human fossils are only about four or at most five million years old. These include a leg bone from the Awash region of Ethiopia, where Don Johanson and his colleagues have worked for some years, an arm bone from the east side of Lake Turkana, and various teeth, jaws, and a wrist bone from the same area. It is clear from these fossil bones that the creatures to which they belonged had already evolved a significant degree of bipedalism. This should not surprise us, given the amount of time that probably separates them (four million years ago) from the origin of the human line (perhaps 7.5 million years ago).

Why haven't we found human fossils older than four to five million years? Mostly because the geological record of this time range in Africa has not been very kind to us. Many people imagine that fossil hunters can go off and look at any time slot they choose. Alas, it is not like that. We can search only in the sediments—slices of time past—that the capricious forces of erosion have exposed. And it just so happens that there are few such exposures that give us the glimpse we need. One day I hope we will find more.

When we do find favorable exposures, I'm confident that we shall be able to recognize our earliest ancestors. Fossil-hunting anthropologists mostly identify early humans by their fossilized teeth, because the hard teeth survive the fossilization process better than other parts of the skeleton. Fortunately, early human dentition is characteristic, although, as we saw earlier, one can sometimes be misled. In the case of the very earliest humans, the very first bipedal apes, however, I suspect that we will not easily recognize them by their teeth, because they may well be much like those of other apes. The first humans may indeed have been indistinguishable from apes, except that they walked on two legs, not four. We shall have to recognize them by their anatomical adaption to bipedalism, specifically in their legs, pelvis, and arms.

The evolutionary shift from quadrupedalism to bipedalism would have required an extensive remodeling of the ape's bone and muscle architecture and of the overall proportion in the lower half of the body. Mechanisms of gait are different, mechanics of balance are different, functions of major muscles are

83

different—an entire functional complex had to be transformed for efficient bipedalism to be possible. That this transformation occurred at all indicates to me two things: first, the pressure for change through natural selection was keen; and second, the transformation itself was, on the evolutionary time scale, rapid. It is the second of these, I believe, that will enable us to identify the earliest human fossils without too much trouble. We will be searching for a bipedal ape.

Can we say what events in prehistory favored the evolution of a bipedal ape? Over the years several hypotheses have been proffered, many of them invoking "modern" human qualities, such as tool making, hunting, and culture. And many of them, as we saw, locate the event on the savannah. How do we separate the possibles from the improbables?

We start with the ecological context. During the past twenty-five million years the global climate has cooled considerably, with an average reduction in temperature of some 20 degrees and a shift in vegetation patterns, including a narrowing of the equatorial belt. More important, Africa experienced other climate changes during this time, changes directly driven by geological events in the continent, specifically in the eastern half.

The principal changes were the circumstances surrounding the opening of the Great Rift Valley, beginning about twenty million years ago. As a result of the separation of tectonic plates running roughly north-south underneath the eastern part of the continent, upwelling lava gradually caused the crust to bulge unevenly, building the Kenyan dome and the Ethiopian dome, each reaching about nine thousand feet above sea level. Like huge blisters on the continental skin, these two domes brought large-scale topography to East Africa. At the time, a swath of dense rain forest stretched across the continent, from the Atlantic coast to the Indian Ocean, home to an increasing diversity of ape species. As the two great domes grew, the patterns of rainfall to the east were disrupted, the result of a growing rain shadow. The eastern forests began to fragment, and patches of open country developed, producing a mosaic of environments, from forest to woodland to shrub and grassland.

When, eventually, there was massive faulting along the line of the tectonic plates, the rift plunged several thousand feet, leaving a deep scar that snaked about three thousand miles from the Red Sea to Mozambique. This deep, meandering valley created yet more ecological barriers and microenvironments. There was constant change, an instability, and through time the environmental mosaic became yet more richly patterned. From frigid highland to parched valley floor, from lush mountain forest to open savannah woodland, from lakes and rivers to acrid volcanic slopes, the range of environments was—and still is—breathtaking. A tremendous diversity of species evolved in Africa: it acted as an engine of evolution. This, I believe, is the key to the origin of the human family. We are one example of the many new species that came about as a result of a dynamic environment, in which, for hominids at least, the highlands were particularly important.

The images we all have of the great plains of Africa, darkened by huge migrating herds, are indeed dramatic. So powerful are they that we tend to project them into the past, thinking that the landscape must always have been like that. Once again, it is all too easy to allow the power of present images to distort our pictures of the past. There's no doubt that the images of the plains intruded themselves into traditional ideas of human origins: our ancestors striding out onto the open savannah, there to become noble hunters. In fact, the great plains and the immense herds on them are relatively recent aspects of the African environment, much more recent than the origin of the human family. For that event—our origin—we must look to the highlands created by uplift and rifting, patchwork environments that offered optimum conditions in which new species could evolve.

By about ten million years ago, with the new topography of the eastern part of the continent still building, there was a great diversity of ape species, something of which we are only now becoming aware. Now there are only three ape species in Africa —the common chimpanzee, the pygmy chimpanzee, and the gorilla. Back then, there were as many as twenty. Between ten million and five million years ago, that wonderful diversity began to decline, in part because of competition from a rising number of Old World monkey species, and in part because of the changing habitat. One of the ape species underwent the dramatic evolutionary transformation to become the first bipedal ape, which, as far as we know, was the only time this mode of locomotion had evolved among primates. As a result, while diversity of most ape groups was declining, among the bipedal apes it suddenly burgeoned. A new evolutionary group began to grow, variations on an evolutionary novelty. The immediate question is how this new diversity of bipedal apes thrived where other apes apparently could not.

We have already dismissed the beginnings of stone tool making as the provider of that evolutionary edge: it appeared far too late in our family's history to have been important in the establishment of the family. And the transformation into a hunting ape can equally be set aside as an explanation, for the same reason. A very "human-oriented" explanation, the hunting ape hypothesis finds no support in the archeological evidence, which indicates that hunting became important in the human career relatively late, beginning, probably, with the *Homo* lineage. No, to discover the nature of that evolutionary edge we have to look for more basic reasons, basic biology, not aspects of human culture. Of the hypotheses advanced in recent years in this category, two seem to me to be interesting. One was proposed by Owen Lovejoy, the other by Peter Rodman and Henry Mc-

Henry of the University of California at Davis. Lovejoy's hypothesis enjoyed a blaze of publicity. Rodman and McHenry's did not. It is easy to see why.

Lovejoy is a fine anatomist, a specialist in the mechanics and origin of bipedalism. He decided some years ago to see whether the biological differences between apes and humans may have provided a competitive edge for the development of upright locomotion. His opening premise was very direct: "Hominids became bipeds for some specific biological reason," he explained recently. "It wasn't for locomotion, because bipedalism is a lousy way of getting around. It must have been for carrying things." As bipeds, humans are embarrassingly slow on foot and not particularly agile in the trees. It is also true that the hands, freed from locomotor duties, could carry things. There are two ways to view this, of course. First, that hominids became bipedal *in order to* free the hands for carrying things. Second, that in becoming bipedal for some other reason, hominids were *able* to carry things. Lovejoy prefers the first of these two views.

Another part of Lovejoy's argument is that, because so drastic an anatomical rebuilding is required to transform a quadruped into a biped, an animal in which the evolutionary change is still incomplete would be an inefficient biped. "During this period, a reproductive advantage must have fallen to those in each generation that walked more frequently in bipedal posture *despite* their lack of efficiency," he reasoned. "When a trait proves to have so powerful a selective advantage, it almost always has some direct bearing on the rate of reproduction. But how could our ancestors have had more successful offspring by walking upright?"

The subsequent argument runs something like this. Apes reproduce slowly, with births spaced widely apart, once every four years. If the bipedal ape could increase its reproductive output, producing infants more frequently, it would be at an overall selective advantage. Because much of biology is driven by energy supply, an increased reproductive output would require greater

87

energy in the female. How was she to achieve this? "Males represent an untapped pool of reproductive energy," concluded Lovejoy. "If a male provisions a female with food, the female has more energy available for parenting, and more offspring can be produced." In order to provision a female, a male must be able to gather food and bring it back to her. Hence the need to walk bipedally, freeing the hands to carry things.

But there is more to it than that. Lovejoy's hypothesis seems to explain everything, perhaps too much. For instance, a male would be foolish to provision a female unless he was sure that her offspring were his. There is no genetic advantage to be had by a male if he helps to rear the offspring of another male. A bond between male and female must develop, in which the female stops publicly advertising sexual readiness but remains constantly attractive to her mate.

In developing this line of argument in an article in *Science*, Lovejoy pulled off one of the best concealed jokes in the scientific literature. "Human females are continually sexually receptive," he stated, and, as is required in scientific articles, provided a reference to back it up. "D. C. Johanson, personal communication," it read, a quip that somehow escaped the notice of the usually fastidious *Science* editors. Citing Frank Beach, a group of respondents to Lovejoy's article later pointed out: "No human female is 'continually sexually receptive.' (Any male who entertains this illusion must be a very old man with a short memory or a very young man due for a bitter disappointment.)"

Back to Lovejoy's argument. He pointed out that where monogamous pairings occur in other primate species, the size of the canine teeth in males is much reduced and is similar to that in females. So it is with the monogamous Asian lesser apes, gibbons and siamangs. And so too is it in early humans, claimed Lovejoy. He speculated that brain size could begin to increase, given the tight, protective social network he envisaged.

It was a very complete description, one that not only explained the origin of bipedalism but also accounted for brain-size increase and the advent of the nuclear family. So complete, so inventive was it, that, according to the Harvard primatologist Sarah Hrdy, "it has all the makings of a powerful myth." It certainly captured a lot of public attention, and its publication in a widely respected, sober scientific journal bestowed on it a certain authority. Pick up any anthropology textbook these days, and you will find "the Lovejoy hypothesis" prominently displayed. Yet it was strongly criticized.

For instance, my friend and colleague Glynn Isaac warned, "A general scientific readership that is not acquainted with details of the state of enquiry in the study of human evolution should be aware that a number of assertions incorporated in Lovejoy's argument are in fact uncertain." Adrienne Zihlman, an anthropologist at the University of California at Santa Cruz, was equally concerned: "Lovejoy's views run counter to evidence on primate reproduction and social behavior and misrepresent the behavior of contemporary hunting-gathering peoples." Referring to the notion of a nuclear family deep in human prehistory, Allan Wilson and his colleague Rebecca Cann said, "We caution against interpreting the fossil record in terms of Western cultural standards." In fact, only about 20 percent of human societies are monogamous, and the nuclear family is very much a phenomenon of modern Western civilization.

For me, the most stinging criticism concerns characteristics of monogamous pairs in primates. Lovejoy points out that in such pairs there is little or no difference in the size of canine teeth—there is no canine dimorphism. But it is also true that in monogamous pairs there usually is little difference in male and female size—no body-size dimorphism. And yet one of the things we can infer from the earliest human fossils is that there was considerable body-size dimorphism: males were about twice the size of females, a difference we see in modern gorillas. The body-size dimorphism in primates is always associated with competition among males for access to females, and some kind of polygyny, with one male controlling sexual access to several females. It never occurs in monogamous species, where one male has access to only one female. So, although Lovejoy's hypothesis is attractive, it seems to trip over the very rules of biology that inspired it. I applaud the attempt, but I think it fails.

The second biology-based hypothesis for the origin of bipedalism is very different from Lovejoy's. For a start, it focuses on locomotion, not the ability to carry things, as the immediate benefit. And it explains only bipedalism, not a whole host of other human characteristics. In this hypothesis, the freeing of the hands is a consequence, not a cause, of bipedalism.

Peter Rodman is a primatologist and Henry McHenry is an anthropologist. Their offices are separated by a few steps along the corridor in the main biology building at the Davis campus. They decided to look at bipedalism from the viewpoint of an ape, asking what it could and could not do. And they rounded up some data on the energetics of walking, in humans and apes, work that had been done some years earlier by researchers at Harvard. "We looked at the data and saw that, for a chimpanzee, walking quadrupedally was no more and no less energetically expensive than walking bipedally," explains Rodman. "So if you imagine that hominids evolved from some kind of quadrupedal ape, then you see that there's no energy barrier, no energy Rubicon in going from quadrupedalism to bipedalism. But the most important point—a new one, as far as we know—is that bipedalism in humans is considerably more efficient than quadrupedalism in living apes."

Previously, people who studied walking in this context had compared human bipedalism with quadrupedalism in conventional quadrupeds, such as dogs and horses. Humans came out second in any measure of the efficiency of energy used for locomotion. But as Rodman and McHenry point out, humans evolved from apes, not from dogs. Not a novel observation, but one that has been overlooked in these kinds of calculations. Chimpanzees are not particularly good quadrupeds energetically, especially over long distances, because their style of locomotion is a compromise between walking on the ground and climbing in the trees.

"If you're an ape, and you find yourself in ecological circumstances where a more efficient mode of locomotion would be advantageous, the evolution of bipedalism is a likely outcome," says Rodman. "What kind of ecological circumstance might that be?" Rodman and McHenry point to the fragmentation of forest cover occurring east of the Rift Valley ten million years ago. As time went on, "food was more dispersed and demanded more travel to harvest." In other words, there was no change in diet other than that the food itself—on trees and bushes—was widely scattered in open woodland rather than densely packed as in the original forest cover. "Bipedalism provided the possibility of improved efficiency of travel with modification only of hind limbs while leaving the [ape] structure of the forelimbs free for arboreal feeding." So, they conclude, "the primary adaptation of the Hominidae is an ape's way of living where an ape could not live."

If true, it means that the first human was indeed simply a bipedal ape. The changes in teeth and jaw structure that we associate with human fossils may have developed later as further environmental change encouraged a gradual shift in diet. We can never be sure, of course, which hypothesis is correct, because, as with all of evolution, we are dealing with a singular historic event. We have to make judgments on what appears to be the most scientifically persuasive. To me, Rodman and McHenry's hypothesis is one of the most persuasive we have. As Sarah Hrdy observes, "Rodman and McHenry's hypothesis is practical, scarcely the stuff of myths."

. . .

I said earlier that the fundamental distinction between humans and apes is that we stand upright, with our upper limbs free. And yet I have suggested that the first human was a bipedal ape. Although this may seem contradictory, it is not. The two statements are based on different perspectives: one, that of history as we know it to have unfolded; the second, that of the biology of the first human. Unless our ancestors had upper limbs free from locomotion, they would not have been able to evolve many of the capabilities that contributed to our humanity, such as the elaboration of material culture within a social context. But the first human species can best be described as a bipedal ape.

In the larger sweep of history, we can look at the changing climate from ten million years ago onward; we can look at the geological stirrings that altered the topography and vegetation patterns of East Africa; and we can say, yes, the first human evolved as a direct response to those changes. The environmental circumstances were favorable for the evolution of bipedalism in a quadrupedal ancestor, the consequences of natural selection. I know that many people would prefer to imagine a more significant beginning to the human family, something a little more awe-inspiring. This sentiment was part of the inspiration of the long-popular myth of the intrepid ape striding out into the savannah, there to triumph over adverse circumstances.

It is also true that some of us believe the very fact of being bipedal somehow bestowed a nobility upon the first human species. By being bipedal, our earliest ancestors undoubtedly gained certain practical benefits, such as the ability to carry things and habitually to have a better view across the terrain. But to assume that they were consequently noble is to view the posture through our own experience as fully modern humans, creatures who have come to dominate the world in so many ways. Having experienced what it is to be human in the fullest sense, we find it difficult, perhaps impossible, to view the world through the eyes of creatures different from ourselves, through the eyes of the first bipedal ape.

So, as the molecular evidence shows, we are closely related genetically to the African apes. When first Goodman and then Sarich and Wilson demonstrated the human–African ape intimacy, the surprise was big enough. But when the relation was measured in numbers, the surprise was even bigger. It turns out that the difference between us, in the basic genetic blueprint, DNA, is less than 2 percent. This is smaller than the genetic difference between a horse and a zebra, which are capable of mating and producing offspring, albeit infertile offspring. There has long been speculation on the possible outcome of sexual union between humans and chimpanzees, fueled in the early years of anthropology by the erroneous notion that apes were a form of regressed humankind. Even in these days of genetic sophistication there are persistent—and always unconfirmed—rumors of "experimental" matings between humans and chimpanzees. As it happens, even though the genetic blueprints of humans and chimpanzees are similar, at some point in human history a change occurred in the packaging of the DNA. The DNA of apes is packaged in 24 pairs of chromosomes, in humans, in 23 pairs, a difference that would probably make barren any such sexual union.

The degree of genetic difference between humans and the African apes is of the magnitude that geneticists usually associate with closely associated, or sister, species. For instance, horses and zebras are placed in the same genus, *Equus*. Yet anthropologists have traditionally placed humans and apes in separate biological families, which implies a big difference indeed. No wonder Morris Goodman wanted to change things in 1962, when he said that humans and apes should be classified within the same biological family.

And now he has a bigger reason than ever to change things, because molecular evidence has just produced potentially the biggest surprise of all. Until recently, the molecular data seemed to indicate that humans were about equally distant genetically from chimpanzees and from gorillas. Chimps and gorillas could be imagined as having split off from the last common ancestor at about the same time, producing the African apes as one group and humans as a second group.

Now, however, more molecular evidence indicates that gorillas may have diverged from the common stock as much as two million years earlier than chimpanzees, some 9.5 million years ago. Chimpanzees and humans separated from each other about 7.5 million years ago. This leaves us with the startling conclusion that the chimp is more closely related to us than it is to the gorilla. Goodman and his colleagues base their conclusion on comparisons of the actual structure—the DNA sequence—of important genes in humans and apes. It is molecular anthropology at its most exquisitely detailed.

"If Morris Goodman is correct in his conclusion, we will just have to go back to the anatomical evidence and find out what we've been missing," says Lawrence Martin, an anthropologist at the State University of New York at Stony Brook. Martin's concern—the concern for many of us—is that chimpanzees and gorillas are similar anatomically, including their unique mode of locomotion, knuckle walking. Knuckle walkers use bent fingers, not the flat of the hand, to support their weight on their forelimbs. "If chimps and gorillas arose separately, it would mean that their similar anatomy, including the knuckle-walking complex, must have evolved independently," says Martin. "Anything is possible theoretically, but it doesn't seem likely."

Martin has recently completed a detailed study of key anatomical features in chimpanzees, gorillas, and humans, looking for signs of relatedness. Jointly with Peter Andrews, a colleague from the Natural History Museum in London, he concludes that,

although the three form a natural biological group, the chimpanzee and gorilla are each other's closest relative, with humans
slightly more distant. "It would be remarkable if this proved not
to be the case," Martin comments. Hence his remark about having to "go back to the anatomical evidence and find out what
we've been missing" if Goodman's latest suggestion turns out to
be correct.

Let us suppose for a moment that this latest conclusion from
molecular biology is correct—and, incidentally, it is not unani-
mously supported by geneticists. Does it have any implications,
other than that we are even more intimately in the camp of the
African apes than we imagined? "It means that it is more likely
than not that the immediate ancestor to humans was a knuckle
walker," suggests David Pilbeam. This addresses Martin's comment about the probability that knuckle walking evolved twice.
"It's more parsimonious to assume that knuckle walking evolved
once only, and was part of the ancestral condition from which
first gorillas and then chimpanzees and humans evolved," David
replies. "In that case, the African apes maintained this ancestral
mode of locomotion, and humans changed theirs."

Does this mean that Sherwood Washburn was right all along,
when he suggested in the 1960s that our ancestors were knuckle
walkers? I can't be certain one way or the other, but I do know
that in the earliest human fossils that could bear vestigial traces
of knuckle walking—the four-million-year-old arm bones from
east Lake Turkana—there is no sign of knuckle walking. There
are strong indications in the wrist's anatomy of an adaptation to
tree climbing, but not to knuckle walking. Perhaps all vestiges
were lost in the time between the origin of hominids and the life
of this individual, a gap of some 2.5 million years, certainly long
enough for a great deal of anatomical evolution. We don't know.
We will only know when we find evidence of the very first bipedal apes. And that, I hope, will be soon.

Whether Goodman and his colleagues are correct in believing the chimpanzee to be our first cousin and the gorilla a more distant second cousin, there is no longer any doubt about our proper place in nature: we are an ape of a rather unusual kind. And Goodman is certainly correct to suggest revision at last of the formal biological classification: the two knuckle walkers and the bipedal ape belong in one family, African apes all.

6

The Human Bush

"*W h a t   d o* you think this means, Walker?" I handed
Alan a Federal Express package from the Institute of Human
Origins, Don Johanson's organization in Berkeley. The package
contained a copy of galley proofs of an article soon to be published in the British journal *Nature*. "New Partial Skeleton of *Homo
Habilis* from Olduvai Gorge, Tanzania," read the title, with Don
and nine others listed as authors. The new fossil was denoted
OH 62, or Olduvai Hominid 62. No covering letter; no explanation. "It means that Don thought you'd been asked to assess the
merits of the manuscript for *Nature*, but it's being published anyway," Alan said with a laugh at his not very subtle joke about the
strained relations between Don and me. As it happens, it was the
first time either Alan or I had set eyes on the paper, but, like
everyone in the profession, we knew it was coming. The anthropological bush telegraph is very effective.

Don and I are the same age, we've both had our share of luck
in hominid fossil hunting, and we've both achieved a degree of

celebrity. Once, we were close colleagues, friends even, and occasionally we sailed together off the Kenyan coast. During the 1970s, when Don and his joint French-American team were making spectacular fossil discoveries in the Hadar region of Ethiopia, we met frequently and talked about the new finds and what they might mean. Don would bring his newly unearthed fossils through Nairobi on the way back to the United States, and we would make accurate fiberglass copies—casts—of them, just in case anything untoward happened on the journey.

About a decade ago our personal and professional relationship began to deteriorate, for reasons I consider best not discussed publicly. One manifestation of the eroding relationship, however, was that, when reporters spotted an opportunity for a "good personality story," they frequently set up Don and me in opposition, often where no real confrontation existed. I attempted to divert such tactics, but it is not clear that others always did the same. In any case, headlines like "Rival Anthropologists Divide on 'Pre-Human' Find" and "Bones and Prima Donnas" began to appear in newspapers and national magazines.

Ostensibly, the "divide" between us was over the interpretation of Don's new fossils from Ethiopia. Don viewed the fossils as indicating a simple pattern for human evolutionary history. He believed that the line leading to us, the genus *Homo*, arose recently. I felt at the time, and still do, that our evolutionary history is probably more complicated than most anthropologists believed, and that *Homo* had much deeper roots than Don's interpretation allowed. Once, in the spring of 1981, I found myself manipulated into a television confrontation with Don over this issue: it was billed as the Leakey line versus the Johanson line, ancient *Homo* versus recent *Homo*. The program was "Cronkite's Universe," filmed at the American Museum of Natural History. After some awkward debate between us, with me privately wishing I were somewhere else, I told Walter Cronkite that, as a

Leakey, I was well used to controversy in anthropology. I then turned to Don and said I thought future discoveries would prove him wrong. I believe they have; ironically, Don's OH 62 find played a significant role in the story.

When I received that Federal Express package from Don, in April 1987, I quickly read through the paper before giving it to Alan and immediately formed a strong impression. "Well, what do you think, Walker?" I asked as he leafed through the photocopied pages. I didn't have to wait long for his response. "That's not *habilis*," he said firmly. "It's much too small." My feeling exactly. Don's paper argued that the partial skeleton he and his colleagues had recovered from Olduvai Gorge belonged to the species *Homo habilis*, the first known species in the line that led to modern humans. Alan and I quickly concluded from the details in the paper before us that Don had made a mistake. "Let's go to lunch and talk about it," I suggested.

In the quest to understand human origins, anthropologists focus on two aspects of prehistory. The first is the overall evolutionary pattern of the human family, such as when new species arise and others go extinct. Often, this is simply referred to as the family tree, the pattern of the existence of species through time. The second concern is the biology of the various species, such as how they subsisted as social groups, what they ate, and how they interacted with other species, including other species of hominids. Obviously, there is a great deal more speculation involved in the second area than in the first, particularly in the realm of species' interactions. In this chapter we shall focus on evolutionary pattern, the shape of the family tree throughout our history, particularly the early part of our history, where most of the uncertainty lies.

We have a good idea of the hominid players who appeared during this early span, between four million and one million years ago. The evidence comes from more than six decades of persistent searches in various parts of Africa.

Before 1925 all the fossil evidence of human history had come from Europe and Asia; it was primarily Neanderthal and *Homo erectus* (known then as *Pithecanthropus*). Africa was considered by most experts to have had little or nothing important to do with human origins, so when Raymond Dart announced in *Nature*, in February 1925, that he had discovered in South Africa an apelike creature that was ancestral to humans, he was rudely scorned. His discovery, from the debris of Taung, a limestone quarry on the southwest edge of the Transvaal, was the fossilized face and part of the cranium of a young individual to which he gave the scientific name *Australopithecus africanus*, or southern ape from Africa. This jewel of a fossil is, however, better known as the Taung child.

Dart recognized that, along with many apelike features, the Taung child also had telltale signs of hominid features. Most important was the foramen magnum, through which the spinal cord leaves the brain and enters the vertebral column. This important cranial landmark was positioned under the middle of the cranium, as it is in humans, not toward the back, as in apes. The Taung child, Dart correctly reasoned, had been a biped; hence, it must have been a hominid. The humanlike structure of the teeth and certain aspects of the brain structure (inferred from a natural endocast from inside the skull) strengthened Dart's conclusion.

Apes have a characteristic dental pattern. The incisors tend to be large and projecting; the canines are often long and sharp, like daggers; the cheek teeth have high cusps, adapted to processing leaves and fruit. In hominids, the incisors are small, as are the canines, and the cheek teeth are relatively flat, adapted to grinding food. The pattern Dart saw in his little fossil closely fitted the hominid pattern. The Taung child was the first early hominid to have been found in Africa; indeed, it was the earliest hominid found anywhere.

However, the universal prejudice that Asia, not Africa, was the site of human evolution, and a barely suppressed revulsion that anything so apelike could possibly have anything to do with our heritage, barred the Taung child from acceptance into the family of man for more than two decades. Finally, the anthropological community acknowledged that Dart had been right: the Taung child was part of our heritage. This led, of course, to a recognition that Darwin was prescient when he concluded in 1871 that Africa had been the cradle of mankind.

Despite the unenthusiastic reception accorded to the Taung child, Dart resumed his search for early human fossils, joined energetically by the Scottish paleontologist Robert Broom. These two remarkable men unearthed many hominid fossils between the 1930s and 1950s from four major cave sites in South Africa. (Ironically, nothing more was ever found in the Taung quarry.) Some of the fossils were like the Taung child, that is, a mixture of ape and human characteristics, in which the lower jaw and cheek teeth were large but not enormous. In others, the same mixture of ape and human characteristics was accompanied by a lower jaw and cheek teeth of huge proportions. The molar teeth were like flat millstones, five times the surface area of modern human molars.

Here were two kinds of bipedal ape, the principal difference being in the robusticity of the lower jaw and the size of the cheek teeth. For a while there was a plethora of scientific names attached to different specimens. Later this nomenclatural confusion was refined and there were just two species' names. *Australopithecus africanus*, the name that Dart had given to the Taung child, was applied to the group of specimens with less massive jaw and teeth. The second species, appropriately, was called *Australopithecus robustus*. Although it was difficult to determine accurate dates for these fossils from the mélange of deposits in caves, it seemed reasonable that *africanus* predated and was ancestral to *robustus*. The *africanus* species was also seen as a likely ancestor to the line that led to us, *Homo*. This, then, was a simple, Y-shaped pattern: a single ancestral species, *africanus*, and two descendant lines, *robustus* on one side and *Homo* on the other.

Then in July 1959, after almost three decades of searching for early human fossils in Olduvai Gorge, my mother uncovered the famous *Zinjanthropus* skull, which caught worldwide attention and assured Louis and Mary of further research funding. *Zinjanthropus* was like *Australopithecus robustus* in having ape and human features in its cranium, with an enormous lower jaw and cheek teeth like millstones. But it was even more robust, an exaggeration of the *robustus* pattern. The skull quickly acquired the sobriquet Nutcracker man. Eventually *Zinjanthropus* was given the scientific name *Australopithecus boisei*, which many considered to be a geographical variant of the South African robust species. The significance of *Zinjanthropus* in the annals of paleoanthropology, however, is that it was the first early human fossil to be found in East Africa.

My parents didn't have to wait long for their next major discovery, *Homo habilis*, the first fragments of which my brother Jonathan found in 1960. During the next three years Louis and Mary and their helpers unearthed many fossil fragments of what they suspected was the same species. (These discoveries included hand and foot bones, rare treasures in the hominid fossil record.) Very similar to *Australopithecus* in many ways, the new species had a less projecting face, smaller teeth, and, most significant of all, a larger brain. *Homo habilis* provided tangible evidence of that second tine of the Y-shaped pattern.

The novel adaptation in the human family was, of course, our mode of locomotion: all human species are variants on the theme of bipedal apes. Dart's, Broom's, and my parents' discoveries showed that these variants are basically of two forms. At one extreme are small-brained creatures with large cheek teeth: the *Australopithecus* species. At the other are large-brained creatures with small cheek teeth: the *Homo* species. These ultra-cryptic descriptions set the outer limits of what we see in the fossil record. The australopithecines have every indication of being fully committed vegetarians, processing large amounts of plant foods between their large cheek teeth. The *Homo* species appear to have been much more omnivorous, and included meat in their diet. In both cases the face protruded more than it does in modern humans, but by no means as far as in modern apes. They were all bipeds, like modern humans, but probably were still adept tree climbers. Until recently, the overall body proportions and build of all the early hominids—*Australopithecus* and *Homo*—was considered similar.

When I initiated explorations east of Lake Turkana, first in 1968, then in earnest the following year and for another decade, something of the pattern of human evolutionary history had therefore been established. *Australopithecus* lived at the same time, and possibly in the same terrain, as *Homo*. The discoveries at Lake Turkana confirmed this. But the pattern was sketchy at best and lacked the clear dimensions of time. The oldest unequivocal specimens we had were not much more than two million years, and we needed to know what came earlier, how the pattern formed itself.

Then, in the mid-1970s, two remarkable discoveries enabled anthropologists to probe farther into our past. First, at Laetoli, twenty miles southwest of Olduvai Gorge, a rare concatenation of geological circumstances preserved a set of hominid footprints close to 3.6 million years old. My mother had visited the site many years earlier with my father, but only in the mid-1970s did she initiate significant work there. What a paleontological treasure had been awaiting her! Three hominids had walked in a northerly direction, their feet leaving crisp prints in newly fallen volcanic ash, debris from the nearby Sadiman, a volcano gently erupting.

Less dramatic than the footprint trail, but more informative in our quest for pattern, was the discovery of a dozen or so

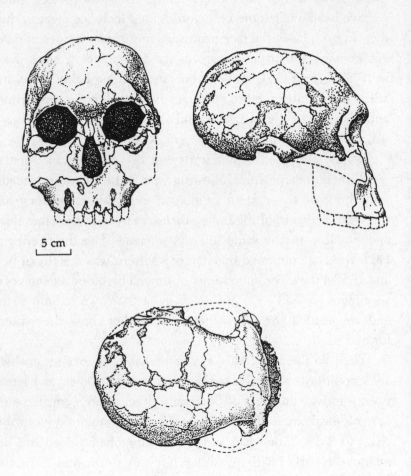

5 cm

Homo habilis is the first known hominid species in which brain expansion has oc-curred. Here, in skull 1470 (found on the east side of Lake Turkana in 1972), the more rounded cranium is very striking. (Courtesy of A. Walker and Richard Leakey/Scien-tific American, 1978, all rights reserved.)

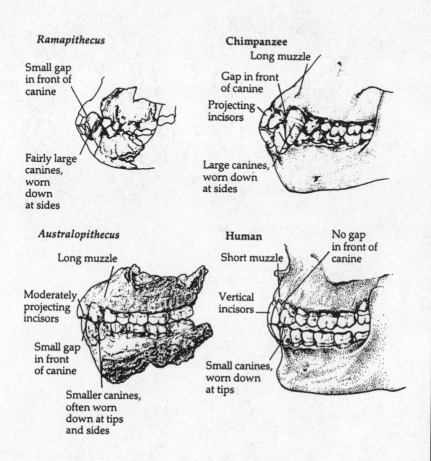

Ramapithecus

Small gap in front of canine

Fairly large canines, worn down at sides

Chimpanzee

Long muzzle

Gap in front of canine

Projecting incisors

Large canines, worn down at sides

Australopithecus

Long muzzle

Moderately projecting incisors

Small gap in front of canine

Smaller canines, often worn down at tips and sides

Human

No gap in front of canine

Short muzzle

Vertical incisors

Small canines, worn down at tips

The tooth characteristics of apes and humans are very different. Here we see Ramapithecus (a fossil ape) and a chimpanzee (a living ape), in which the jaw projects forward, the canine teeth are large, and a gap (diastema) exists between the upper canine and incisor. In humans the jaw does not project forward, the canines are small, and there is no diastema. Australopithecus, an early member of the human family, is somewhat intermediate: although the canines are relatively small, the jaw projects and (in some cases) a diastema exists. (Courtesy of the British Museum [Natural History].)

105

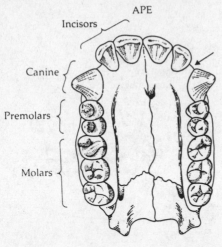

APE

Incisors

Canine

Premolars

Molars

Chimpanzee upper jaw

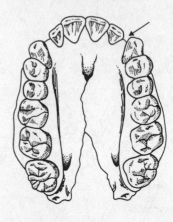

LAETOLI-HADAR
Dental arcade and diastema

A. *afarensis* upper jaw (AL200)

HOMINID
(*Australopithecus* and *Homo*)

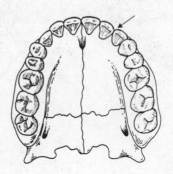

Human upper jaw

The very different shapes of the ape and human jaws are obvious here, with the ape's being substantially U-shaped and the human's arched. The Laetoli-Hadar specimens (often called Australopithecus afarensis*) has something of an ape shape. Arrows point to the diastemas in the afarensis and chimpanzee, and the position, now closed, in humans. (Courtesy of Luba Gutz.)*

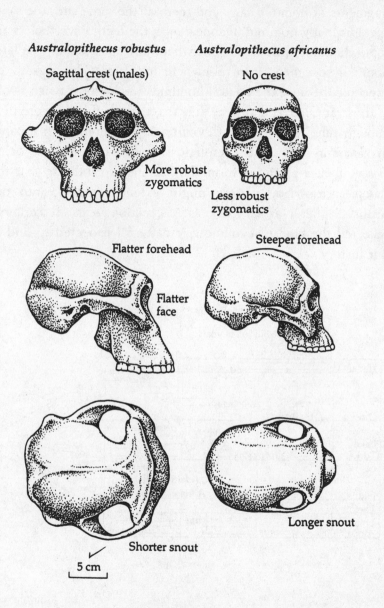

Australopithecus robustus

Sagittal crest (males)

More robust zygomatics

Flatter forehead

Flatter face

Shorter snout

5 cm

Australopithecus africanus

No crest

Less robust zygomatics

Steeper forehead

Longer snout

The robust australopithecines (including Australopithecus robustus and boisei) differ from the gracile species (Australopithecus africanus) mainly in dental and cranial features, not overall body structure.

107

fragments of hominid jaws and teeth at the same site, the same age. Distinctly hominid in appearance, the teeth nevertheless are somewhat more primitive, somewhat more apelike than the later hominid specimens from elsewhere in Africa. This was to be expected, because the Laetoli hominids were almost twice as old as those at Olduvai and Lake Turkana; there had been more than enough time for significant evolutionary difference. In a paper in *Nature* in 1976, the Laetoli fossils were identified as being *Homo*, albeit a primitive form. This seemed to indicate that the *Australopithecus-Homo* pattern might extend far back into our history, at least as far back as 3.6 million years. It certainly reflected the kind of evolutionary pattern I expected to find in our history.

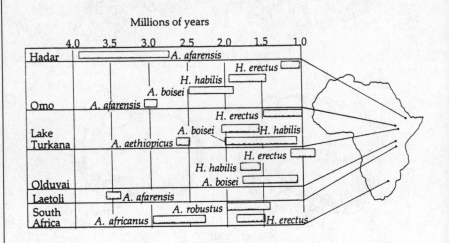

All the earliest hominid discoveries have been in Africa, confirming Charles Darwin's prediction that Africa would be shown to be the "Cradle of Mankind." The major sites of discovery have been in South Africa, Tanzania, and Ethiopia, as shown here.

Just as the Laetoli hominid jaws and teeth were published in the scientific press, Don Johanson and his colleagues announced their first assessment of the hominid fossils they had been unearthing in Ethiopia since 1973, many of which I had seen in Nairobi. These included the famous Lucy partial skeleton and the so-called First Family, a large collection of fossil fragments from perhaps fifteen different individuals. Close to three million years old, these fossils were said to belong variously to a primitive *Homo* species and one, perhaps two, *Australopithecus* species. Again, the Y-shaped pattern seemed to be present: *Australopithecus* on one hand, *Homo* on the other.

We have no clear picture of what actually happened earlier than about four million years ago, back to the origin of the hominid family, because we lack fossil evidence. We have yet to find fossil-rich deposits that will allow us to sample this slice of our history. But if we assume that the hominid family did indeed first appear about 7.5 million years ago, we can assume that soon afterward other hominid species would have evolved, descendants of the founding species, variants on the theme of the bipedal ape. Hominids, like other large terrestrial species, would have embarked on what biologists call an adaptive radiation.

Adaptive radiations are the norm in evolution. A species evolves a novel adaptation—in the case of humans, it was upright walking—and effectively becomes the founding member of a new evolutionary experiment. Very soon new species arise from this founding member, and a cluster of new descendant species appears through time. For example, the *Alcelaphini*, a tribe of African antelope that includes the blesbok, hartebeest, and wildebeest, became extremely effective and successful grazing machines, living on tough forage, and spread over much of sub-Saharan Africa. The tribe first appeared a little more than five million years ago, represented by one species, and now has ten existing branches in its evolutionary bush. In shape the evolutionary history of the *Alcelaphini* tribe looks like a flat-topped acacia tree.

What of humans? The shape of our evolutionary bush as I've presented it so far is much more simple. Again, it began with a single species, the trunk of the tree. The tree then bushed out, with two main branches, *Australopithecus* and *Homo* species. Then, toward the present, the *Australopithecus* branch got pruned off as the australopithecine species became extinct. Ultimately, only one branch remained at the tip of the tree: us, *Homo sapiens*. The initial adaptive radiation, followed by a dramatic winnowing to a single remaining species, is the pattern of the human bush. But, as a Y-shape, with two main branches, it seems too simple, somehow incomplete.

My father didn't much like this kind of pattern, because he didn't believe that *Homo* had descended from *Australopithecus*—any species of *Australopithecus*. I did not care for it either, because when I began expeditions to Lake Turkana in the late 1960s I saw signs of a complexity in the human bush that indicated roots deeper than an *africanus* ancestor. I felt that future discoveries from there earlier in our history would fill out the complexity. Species from the period between three and two million years ago in our history would give our tree a more bushy appearance than the Y shape. But how many more species we might find, I could not predict.

When he delivered the Huxley Memorial Lecture in 1958, titled "Bones of Contention," the eminent British anthropologist Sir Wilfrid Le Gros Clark remarked, "Every discovery of a fossil relic which appears to throw light on connecting links in man's ancestry always has, and always will, arouse controversy." His observation applies doubly for *Homo habilis*, which marks a key juncture, a connecting link, in human ancestry. *Habilis* has always been controversial in anthropology. It was for my father. It was for me. And, from a glance at the OH 62 paper he sent me, I knew it was going to be so for Don.

"You know, they call this a partial skeleton, but it's really scrappy, I mean *really* scrappy," said Alan as we installed ourselves at a little Italian restaurant five minutes from the museum. The list of fossil fragments was long, a total of 302 pieces in all, parts of the skull, right arm bone, and both legs, but frustratingly incomplete and badly eroded. The creature was described as being extremely small, about three feet in height, with long arms and relatively short legs. "When you look at this, and think about some of the things we've got—like 1500, 3735—you wonder what all this fuss is about," I said, my competitive instincts rising. (Numbers like 1500 and 3735 are museum accession codes for the fossil specimens we have found, and conversations among anthropologists can seem somewhat arcane to outsiders, because they are often replete with such numbers and devoid of names.) "When we get back to the museum, we'll dig out some of our 'partial skeletons,'" I said. "I think we can have some fun." We planned to make comparisons among some fossil specimens, and we expected to see something interesting.

We also knew that the episode unfolding before us was certain to revive an old problem of *Homo habilis*, one that has dogged anthropologists since the species was discovered and has impeded attempts to reveal the nature of the evolutionary pattern we were seeking. When, in April 1964, my father, in company with Phillip Tobias and John Napier, announced in the pages of *Nature* the discovery of *Homo habilis*, a new species of hominid they described as the maker of stone tools and the ultimate ancestor of modern humans, the response was immediate, loud, and highly critical. The critics, incidentally, were led by none other than Le Gros Clark, thus fulfilling his "Bones of Contention" statement of six years earlier.

One reason Louis and his two colleagues had opprobrium heaped on their heads so liberally was that, in making their new fossil a member of the genus *Homo*, they were forced to modify the definition of *Homo*. A second was that the collection of fossils brought together under the name *Homo habilis* was highly variable anatomically, too variable to represent just one species, according to many anthropologists.

Eventually the decibels of the debate died down, and *Homo habilis* became accepted as a valid species, the immediate precursor to *Homo erectus*. However, the issue of extreme anatomical variability in the species sample persisted, and debate on this remains as vigorous as ever, leading one prominent anthropologist recently to publish a paper with the title "The Credibility of *Homo habilis*." This gives a sense of the uncertainty anthropologists feel when they speak about the issue.

"Well, it depends on which specimens you feel you'd like to include" is the typical response when someone is asked for an opinion on *Homo habilis*. Of the several dozen specimens that have been said at one time or another to belong in this species, at least half probably don't. But there is no consensus as to which 50 percent should be excluded. No one anthropologist's 50 percent is quite the same as another's. It was into this paleontological mire—what qualifies as *Homo habilis* and what does not—that Don and his colleagues inserted themselves with their conclusions about their diminutive new Olduvai fossil, OH 62.

Something of a sense of *déjà vu* struck me about this; a decade and a half earlier, I had been down a similar path. Three years after I led the first major expedition to the eastern shore of Lake Turkana, skull 1470 was found, a fossil that did for me what Zinj had done for Louis: it made me famous, put me on the international stage. That was in August 1972, three months before Louis died. The cranium that goes by the now famous accession number 1470 was clearly of the large-brain, small-cheek-teeth type of hominid. With a cranial capacity of close to 800 cubic centimeters, 1470 was obviously a good candidate for *Homo habilis*, as my colleagues repeatedly told me. However, I insisted on publishing it as *Homo sp.*, meaning, yes I agree it is *Homo*, but I'm not prepared to say what species it is. The *Homo habilis* affair was still so messy at the time, I thought it appropriate to leave the fossil in a "suspense account."

For my caution I was roundly criticized. Several colleagues suggested that I do the obvious thing and call it *Homo habilis* or have the courage to name a new species. I refused to do either, and *Homo habilis* is what 1470 generally came to be called, very much by default. These days I believe that this is probably correct, but only if OH 62 is *not* accepted as *Homo habilis*. Let me explain how this conclusion unfolds.

By the time 1470 turned up at Koobi Fora, we had also found specimens of *Australopithecus boisei*, the small-brain, large-cheek-teeth hominid, like Zinj. These fossils came from almost two million years ago, like the oldest deposits at Olduvai Gorge. Olduvai and Koobi Fora, then, were displaying the same pattern: the two adaptive extremes, one a small-brain, large-cheek-teeth species (*Australopithecus boisei*), the other a large-brain, small-cheek-teeth species (*Homo habilis*). Specimens of *Homo erectus* also turned up at both sites, further filling out the pattern. But then there was fossil number 1813, a cranium discovered at Koobi Fora a year after 1470, and just a bit younger geologically. It had small cheek teeth and a large face, like *Homo*. But its brain was small, like *Australopithecus*. It was an unexpected combination of features, an enigma.

My colleagues argued vigorously over 1813's status. Some were sure it was a female *Homo erectus*. Others said, no, it was a female *Homo habilis*. Still others championed 1813 as an East African *Australopithecus africanus*. I was sure of two things. First, we could not say definitely what 1813 was. And second, here was evidence that the pattern of human evolution—the bush—probably was more complex than most anthropologists were prepared to admit.

Alan and I wrote an article in *Scientific American* in August 1978 in which we addressed the problem of the Koobi Fora fossils and the overall pattern of human evolution. We were pretty certain that 1813 was not *Homo habilis*, yet we noted that its upper dentition bears "a striking resemblance to the teeth of one of the *Homo habilis* specimens from Olduvai, OH 13." We went on to say, "In all parts that can be compared—the palate as well as the teeth, much of the base of the skull, and most of the back of the skull —the two specimens are virtually identical."

This could mean one of two things, of course. Either we were wrong in thinking that 1813 was not *Homo habilis*. Or many *Homo habilis* supporters were wrong to include OH 13 in their population of specimens. We stuck our necks out and guessed that 1813 was not *Homo habilis*, and that it would prove to be an unidentified small hominid species. This interpretation would give us three separate branches in our tree of two million years ago: *Homo habilis*, the robust *Australopithecus*, and whatever 1813 and OH 13 were. What had been recovered from the Hadar region of Ethiopia made it likely that this greater complexity ranged back at least as far as three million years, a satisfying development from my point of view.

Six months after our *Scientific American* article appeared, Don Johanson and Tim White published one of the most talked-about anthropological papers in recent times. The paper, in the January 26, 1979, issue of *Science*, seemed to sweep aside our analysis and offer a wholly different evolutionary history for humans. The new pattern had been inspired by Don's reassessment of the fossils from the Hadar, including the famous Lucy partial skeleton. "All previous theories of the lineage leading to modern man must now be totally revised" was how *The Times* of London greeted the family tree proposed by Don and Tim.

When I had seen many of the Ethiopian fossils in Nairobi, during those trips Don made on his way back to the States in the mid-1970s, I was impressed by the range of size. Some individuals had clearly been more than five feet tall; others, particularly Lucy, were not much more than three feet. There was also a great deal of anatomical variability. These were the reasons Don initially considered that three different species were represented among the fossils, a large and a small *Australopithecus* and an early *Homo*. He had announced this in his March 1976 *Nature* paper, with his French colleague Maurice Taieb. I agreed with him.

Now, Don had changed his mind, the result of the reassessment of the fossils with Tim White, an anthropologist at Berkeley. They concluded that the Hadar fossils represented just one variable species, not three separate species. Furthermore, they said that fossil teeth and jaws from my mother's footprint site at Laetoli were from exactly the same species as at the Hadar. The fact that the Laetoli fossils were located a thousand miles to the south and were half a million years older apparently did not worry Don and Tim. Theirs was an audacious move that raised many eyebrows among evolutionary biologists.

Don and Tim said that the Hadar and Laetoli fossils were all from the newly named *Australopithecus afarensis*. Furthermore, they argued that *afarensis* was the ancestral stock from which all hominid species evolved. Tim described the species as follows: "Its brain was small, its canine teeth large, and its other teeth were primitive in many aspects. The shape of the dental arches and skull and the protruding face were more apelike than humanlike. Males were substantially larger than females—a condition similar to that seen in modern African apes."

In their new version of the human bush, Don and Tim considered *afarensis* the trunk, leading in one direction to *africanus* and to *robustus-boisei*, and in the other direction to *Homo habilis* to *erectus* to *sapiens*. Again, a Y-shaped pattern. Notice, there is no room for an 1813–OH 13 type of hominid, a third branch in the bush. Much too simple a pattern for my taste. "We considered this scheme and thought it less likely than another," Alan and I wrote in a letter to *Science*. We went on to explain that, because we saw three human species in the two-million-year-ago time slice, we thought it probable that there was more than a single species extant when Lucy lived, less than a million years earlier.

"We recognized two [at two million years ago] in our assessment," retorted Don and Tim. They seemed to be locking themselves into the scheme that would effectively shape their interpretation a decade and a half later, of OH 62. Even though they also said, in their response to our letter, "We realize that all phylogenetic hypotheses require modification with new evidence," they were later curiously resistant to such modification.

The anthropological community threw itself into a turmoil over Don and Tim's *Science* paper, some supporting the validity of *afarensis*, some challenging it. A decade of debate convinced most anthropologists that Don and Tim had, after all, been correct. This included Alan, but not me. I have held tenaciously to the minority position over the years, and an announcement from Don's Institute of Human Origins in March 1991 signals to me that I may soon no longer be quite so alone. After a decade and a half of being unable to conduct field work in Ethiopia, Don and his colleagues returned in 1990, and found yet more fossils from the same time range as Lucy and the First Family. "The majority of specimens further document the presence of *Australopithecus afarensis* at Hadar," said the announcement. "However, some of the finds exhibit anatomical characteristics not previously observed in the Hadar hominid collection." The announcement goes on to say that the new discovery "may rekindle arguments that *A. afarensis* actually consists of more than one species."

I have not seen the new specimens. Few people have. But I would be surprised if they do not become the basis for another reassessment of the Hadar hominids. And I expect that several hominid species will be seen to have occupied the Hadar region some three million years ago, just as the available evidence shows they did at Lake Turkana a million years later.

One thing that has emerged from all this debate, and on which all anthropologists seem to agree, is that in the early hominid species—the *Australopithecus* species—the males are as much as twice the size of the females. In *Homo erectus*, just as in modern humans, there is not much difference between males and females, males being some 10 to 20 percent bigger. Clearly, an important change occurred between *Australopithecus* and *Homo*, one that probably reflects a dramatic change in human social biology. "Usually it is associated with a reduction in competition between mates, and possibly even a significant degree of cooperation," explains Robert Foley, an anthropologist at Cambridge University, about a moderate body-size difference between males and females. This is seen in chimpanzees, in which males are some 20 percent bigger than females, comparable with modern humans. "Chimpanzee social groups are unusual among large primates, in that the mature males in them are often closely related to each other, and they cooperate assiduously with each other, in obtaining mates and in defense against other groups," says Foley.

The important turning point in human history denoted by *Homo erectus* is marked by substantial brain increase, improved technology, use of fire, the development of hunting, and the first hominid migration from Africa; in other words, many of the more "human" aspects of our species. The appearance of human-like body-size dimorphism in *Homo erectus* must be added to that list. But what about *Homo habilis*? Had the reduction in body-size dimorphism already been completed in this species?

For Tim White, the answer was unequivocal: the reduction hadn't even started. "Because *Homo habilis* is considered to be an evolutionary intermediate between relatively small *Australopithecus afarensis* and the relatively large *Homo erectus*, everyone assumed *habilis* would be intermediate in size," Tim said. "People have viewed human evolution through glasses of gradualistic change. Well, OH 62 has smashed those glasses. The change was obviously abrupt, with a big modification in body form between *habilis* and *erectus*."

Could it be, as Don and Tim claimed, that sexual dimorphism in body size in *Homo habilis* was as extreme as it was in *afarensis*, with males being twice the size of females? And that the adoption of humanlike dimorphism happened in one jump, when *Homo erectus* appeared? I think not. The clue is in the arms.

A key measure here is the comparison of the upper arm (humerus) length with the length of the thigh bone (the femur). This gives the so-called humerofemoral index, which in modern humans is about 70 percent. That is, the humerus is only 70 percent of the length of the femur. Modern apes have much longer arms; the humerofemoral index in chimpanzees, for example, is about 100 percent. The humerus is the same length as the thigh bone. Anyone who has been to the zoo has seen how different humans and chimps appear in this respect; the apes' arms dangle below their knees.

What of early humans? In Lucy, the index is 85 percent, somewhere between humans and chimps. Now we come to Don's new fossil, OH 62. "We estimate the OH 62 humerus length at 264 mm," noted the *Nature* paper. This is fully 27 millimeters *longer* than Lucy's humerus, and Lucy was a marginally taller animal than OH 62. As a result, wrote Don and his colleagues, OH 62 has a "humerofemoral index of close to 95%." This figure puts OH 62 closer to chimps than it does to Lucy, and that strikes me as odd. It should have struck Don and his colleagues as odd too.

"They've got themselves into an intellectual knot over this," said Alan, as we drove back to the museum after lunch. "If they have an evolutionary scheme that goes from *afarensis* to *habilis* to *erectus*, then the limb proportions go from less apelike at something under three million years in *afarensis*, to more apelike at 1.8 million in OH 62, and then back to less apelike at 1.6 million in *erectus*." More like an evolutionary hole than an intellectual knot, it seemed to me. "That's not very likely," I suggested. "Not very likely at all," responded Alan. "Come on, let's look at some bones."

The hominid fossils are kept in a strongroom in one corner of the new building at the museum. There are padded tables on which we could spread out the fragile fossils without fear of damaging them. We decided to focus on one fossil in particular, 3735, a 1.9-million-year-old partial skeleton, the first piece of which was found in 1975. It is not a beautiful fossil, but it was precisely right for our purpose, a comparison with OH 62. There were parts of the cranium, a piece of shoulder blade, collarbone, some arm bones, some hand bones, part of the sacrum, and some leg bones. We quickly decided, yes, we would do the same kind of calculations that Don and his colleagues had done with OH 62.

The first thing that was obvious from 3735 was that it was a robustly built creature in its upper body, particularly around the shoulders and arms. The key observation, however, was that it had very long arms, just like OH 62. Again, apelike proportions on a diminutive body. Were we looking at the partial skeleton of a member of the small, 1813–like species? The fact that, as far as we could tell, the cranium of 3735 was small, like 1813's, encouraged us to think so. We suspected that OH 62 also would have had a small cranium, like 1813.

It began to look as if the ideas we'd had in the mid-1970s were correct: around two million years ago, the immediate pre-*erectus* period, there were indeed two nonrobust hominid species, one large, like 1470, effectively *Homo habilis;* the other small, like 1813 (and OH 62), which had surprisingly apelike body proportions and a small brain. But we needed some indication of the body size of the "real" *Homo habilis*, 1470 for our purposes. We got that from two further fossil specimens in our collection, some leg bones and part of a pelvis. Both obviously had belonged to large individuals, like 1470. The pattern was filling out very nicely.

Our examination of some fossils of our own, prompted by Don's claims for OH 62, certainly had clarified matters. We became convinced that the evidence pointed to three separate species—three branches—in the immediate pre-*erectus* period: the robust *Australopithecus* species, *Homo habilis*, and a third species with apelike proportions in its arms and legs. The same pattern probably extended back three million years, perhaps further. "Satisfying, very satisfying," said Alan with a grin when we reached the end of our analysis. I agreed. "Yes, let's have a beer."

So what does the overall pattern of human history look like to us now? We assume that the human bush took root 7.5 million years ago, but there is a lack of fossil evidence between that point and four million years ago. All we can say is that a single species of bipedal ape established the family and probably began an adaptive radiation. We can guess that the bipedal apes in this period were subsisting on apelike diets, but were ranging over larger areas in woodland country, not dense forest and not savannah.

By three million years ago our bipedal apes' teeth were less apelike, so presumably a shift in diet had begun, though it still consisted principally of plant foods. How many branches on the bush there were at this time is, as we've seen, a matter of dispute: my guess is at least two. By two million years ago the bush was visibly spreading, with three or more branches. Some bipedal apes had become small-brain, large-cheek-teeth, specialist plant eaters; others, large-brain, small-cheek-teeth omnivores. Some filled niches in between.

By 1.7 million years—the time of *Homo erectus*—the large-brain, small-cheek-teeth adaptation began to dominate; eventually it was the only occupied bipedal ape niche. We were the occupant. An initiation, a blossoming, and a trimming—that's the pattern of our human bush. It looks to our eyes more complex now than it did a decade ago.

The arrival of the Black Skull made the pattern more complex still.

7

The Black Skull

"*T h i s   t h i n g* is flying like a hippo," I shouted to Alan as the Cessna finally made it over the shoulder of the Rift Valley, thirty miles north of Nairobi. "Maybe we shouldn't have gone back for the meat." That, Alan pointed out, would have made us extremely unpopular. He was right. Kamoya and his crew had been up on the west side of Lake Turkana for two weeks, moving fifteen tons of earth, preparing the ground for our second season at Nariokotome. They needed the supplies that were stashed in every conceivable space in the plane, weighing it down so that it flew uncharacteristically—and uncomfortably—with its nose slightly up. That slowed our speed as we mushed through the air. And we were an hour late, because we had had to make an extra trip to pick up the forgotten meat. No, we couldn't have gone without it.

It was the beginning of August 1985, just a year after we had closed up camp on a season of excavation so remarkable as to be the stuff of dreams: the Turkana boy, virtually a complete skeleton. "What are our chances with the hands and feet, Walker?" I mused.

"Well, an awful lot of dirt's going to be shifted in finding out," he responded. Alan never does rise to a rhetorical question, and this certainly was one. No one could know whether we'd find hands and feet, plus some arm bones, and of course the teeth. They were our goal.

As week after week passed, the excavation work seemed hotter and dustier than the previous year, no doubt because of the huge amount of hill we had moved, opening up a great area of bare earth. And, as we feared, the site was proving to be frustratingly barren. An arm bone, part of a rib—little else. "I don't fancy another season in this awful hole," lamented Alan in his diary. For me a quiet day's excavation, even if fruitless, was a welcome reprieve from the tribulations of museum business back in Nairobi. But, day after fruitless day, it can be wearing.

Almost a month into this frustrating season—on August 29— our *Nature* paper on the Turkana boy was published in London. Celebrations on that account were completely forgotten, however, as the events on that day turned out to be another reminder of how unpredictable this business of ours is. By that time I was back in Nairobi on more museum business.

"It was a day from a *National Geographic* TV special, blazing hot with a cloudless blue sky," was how Pat Shipman later described it in an article in *Discover* magazine. "As we sat on a hilltop, I could envision the broad panoramic sweep of the camera across the miles and miles of treeless badlands, followed by a zoom shot of the dedicated, sunburned scientists [us] plying their craft." Alan's description, written in his diary that evening, was more cryptic: "A crazy day."

The day began with a reprieve from work in the Hole while the crew removed more of the overlying hill in the northeast section. Alan decided this would be a good time to work on a magnificent hippo skull near John Harris's camp, about twenty miles south of Nariokotome and two miles up the Lomekwi River. Pat joined him, while Meave, Mwongella, and David continued excavation of a carnivore skeleton nearby. "The hippo skull was in soft running sand, so excavation was easy," recalls Alan. "By lunch time we had it up on a pedestal and Bedacryled once"—normal routine for excavating fossils. We cut away the ground around the fossil itself, leaving it completely exposed except for the column of earth on which it rests—the pedestal. With delicate fossils, ones that crack when exposed, we apply Bedacryl to harden them. Once the Bedacryl has done its job, we can make a plaster of Paris "jacket" around the fossil, cut through the pedestal, and lift the whole thing out of the ground. In the case of the hippo skull, a second coat of Bedacryl was required, a job for after lunch.

"John came back with us after lunch," explained Alan, "because I wanted to show him a primate humerus that Kamoya had found the week previously. Pat began work on the skull again, and John and I went to look for the humerus. We looked at three spots, but I couldn't remember exactly where it was, so John said he'd walk back to camp and leave the Land Rover with us." If Alan had been the one to discover the primate fossil in the first place, there's no doubt he would have remembered the location. Spatial memory works like that in fossil hunting, especially in this kind of terrain, which to the unfamiliar eye can have a deceptive sameness. But Alan had only been shown the location by Kamoya, so the memory didn't stick.

"On my way back to Pat, I saw another likely place, so I took a short detour, but it wasn't there either," continued Alan. "I gave up, thinking that I'd have to ask Kamoya again. As I walked over a ridge I saw a piece of dark-colored fossil next to a small pile of stones." Whenever a member of the Hominid Gang finds a potentially interesting fossil, he marks the spot with a small cairn, and then checks it out later, with John or Alan or me. As far as

Alan knew, no one had checked this one. He picked it up, and saw that it was a part of upper jaw with enormous tooth roots, so big that for an instant he thought it was some kind of bovid. "Then I saw another piece, from the front of the skull, and thought it was a big monkey. But then I turned it over and saw a frontal sinus—hominid!" Old World monkeys don't have such a system of cavities within the bone above the nose, but humans do.

Alan put the fossils back where he'd found them and walked quickly to where Pat was working on the hippo, about a hundred yards distant. Casually he said, "When you've finished with the Bedacryl, I'll show you a hominid." "Great. What part?" she asked, thinking that John had shown him some scrap, a tooth or piece of jaw. "Oh, skull" was the laconic reply. Alan likes to play the nonchalant Englishman, but the excitement that built over the next half hour had even him grinning widely.

He took Pat to the fossil, and then went to fetch Peter Nzube and the crew. Aila set off in search of John and Meave, and very soon everyone was on hands and knees, carefully scouring the ground around the little cairn. More and more pieces of the dark fossil were found, all fragments of skull, and very interesting they looked, too.

"If I show you a three-million-year-old hominid skull, will you get me a beer?" Alan asked Kamoya, when, dusty but happy, the fossil collectors later got back to the Nariokotome camp. Kamoya had been working at the Turkana boy site all day and was unaware of the new find. He went off to fetch a couple of beers, but assumed Alan had been joking. Alan then gave Kamoya the collecting bag and watched as he carefully opened it. Kamoya quickly realized Alan had indeed been joking, but in a different way. "Look at the size of those tooth roots" was all Kamoya could say when he saw the upper jaw. "Look at the size of those tooth roots." They were as big as anyone could remember seeing on a hominid. Here, it seemed, was one of the small-brain, large-cheek-teeth types of bipedal ape, and at least 2.5 million years old.

"Clearly it's an early robust," Alan wrote in his diary that night. A restrained comment, given that the skull would prove to be one of the most interesting and important specimens discovered in a decade. It would force us to study again—and revise again—the pattern of our evolutionary history.

The radio call came early the next morning, on a bad channel, so that I could hardly hear what Kamoya was saying—or, rather, shouting. I got the salient points, however, and quickly decided that some of those important meetings for which I had returned to Nairobi weren't quite so important after all. If what I'd just heard through the acoustic fog of Kamoya's call turned out to be correct, I could see that Alan's fossil was going to cause quite a fuss.

Meetings canceled, schedule rearranged, I flew up to Nariokotome the following day, Saturday. Because circumstances had made the previous day so hectic, I hadn't had much time to think about the new hominid. But airborne and on my way to the lake, museum business behind me, I begin to muse on the implications of the fossil. As Alan observed later to a newspaper reporter, "It's a lot of fun because it stirs things up just when people were getting complacent."

Although opinions differ concerning the pattern of the human bush, at this time there was virtual unanimity about the history of the robust australopithecines, *Australopithecus robustus* and *Australopithecus boisei*. Both are extreme versions of the small-brain, large-cheek-teeth hominid, with *boisei* the most extreme of all. Populations of *robustus* had lived in South Africa; *boisei* was an East African animal. Geographical variation was the usual explanation advanced for the anatomical differences between them. Virtually all family trees drawn for hominids included a neat progression: from *africanus* to *robustus* to *boisei*, an evolutionary trend through time, in which the chewing apparatus got bigger and bigger.

Alan's new fossil would turn the neat scheme on its head. If hyper-robust anatomy was present right there at the beginning, at least as early as 2.5 million years ago, then such features could not be the end product of *africanus-to-robustus-to-boisei* evolution. Had there been *two* lineages of this type of robust australopithecine, one of them leading to *robustus*, the other to *boisei?* That would make the human bush even bushier, our evolutionary pattern more complicated. To which species did Alan's hominid belong? Was *boisei* itself present 2.5 million years ago, or is the fossil a precursor species? These were the thoughts passing through my mind as I flew up to Lake Turkana that morning, checking off the familiar landmarks on the way, eager to be there, to see for myself what had been found.

When I arrived at Nariokotome Alan was grinning with pleasure, his own and mine. He had already glued together many of the skull fragments. "Like it?" he asked, as if he were showing me a vase he'd bought. "Like it?" I replied. "It's wonderful, fascinating. Wait till they see this," I said, anticipating my colleagues' reaction to it. We were in the mess tent, shaded from the morning sun. I handled the partly reconstructed skull with care, thrilling to the touch of so special a find. Yes, it was an astonishingly robust individual. No doubt about it, the family tree would have to be redrawn. "The new skull is going to force many people to change their minds," I later noted in my diary. "I am even more convinced that the three hominids of 2 million years ago are going to be traced back beyond 3 m.y. Also, it is likely that Johanson-White can be shown to be very wrong in their scheme. We shall see. It's going to be interesting for some."

During the following week we worked at Alan's hominid site every day, and every day we recovered more pieces. It became clear that we were going to get an almost complete skull—the Black Skull, as it came to be known. Manganese salts that penetrated the bone during fossilization had produced a bronze-black color, quite beautiful. On the third day Pat found a big piece of the top of the skull, toward the back. It sported an enormous saggital crest, the bony keel that runs from front to back of the cranium and provides an anchor for the huge muscle that works the jaw. It was the biggest saggital crest I had ever seen on a hominid. Very impressive.

Back in camp Alan was able to piece together very quickly the new pieces. The cranium presented no problem, but the face looked odd, like nothing we'd seen. Neither Alan nor Meave— the wizards of the paleontological jigsaw—was able to get the face bits to fit with the cranium. Very odd. So odd did it look that Alan wrote in his diary, "The skull is not any species of robust we know, and we'll call it *Australopithecus kamoyensis* in Mack's [Kamoya's] honor."

The problem was the orbits, the eye sockets. "Some pieces of bone that obviously belonged to the face had a smooth curve on them," recalled Alan. "I just couldn't get it; Meave couldn't get it; no one could. I almost began to think they didn't belong to the skull, though I knew they must, because there was nothing else at the site. It turns out that the orbital margin on the lateral side has no edge." If you put your index finger on your temple and move it forward toward your eye, you feel a distinct bony edge as you just reach the eye. That's the lateral edge of the orbit. In gorillas, there's no distinct edge, just a smooth curve of bone. This pattern—though less marked—occurs in a few of the robust australopithecines from South Africa, but we had not found it in East Africa.

I hadn't seen Alan stumped on a paleontological reconstruction for a long time. "Don't worry, Walker," I said. "Just keep on trying. You'll get it." With that, I returned to Nairobi, taking Meave, Samira, and Louise with me. Two days later he did get it.

"I had to resort to looking at the tiny slivers of bone that had spalled off the sides of the maxilla," Alan said. "I started gluing them, then putting these onto the side, and suddenly I saw how it went together. All the pieces were there, just split into tiny fragments."

With the breakthrough, and the face fitting snugly onto the cranium, the Black Skull looked much less odd, in spite of the smooth outer margins of the orbits. Nevertheless, the extraordinary creature showed our theories to be far too simple. "No one could have predicted this combination of features" was Henry McHenry's assessment when later he saw the skull. "It turns a lot of our ideas upside down, about the sequence of evolutionary changes in the skull and about who is related to whom."

The unexpected combination of features to which Henry alluded was this. The brain case is very primitive, in its shape and in the arrangement of crests, including the saggital crest. This kind of arrangement occurs in *afarensis* and in the African apes, but not in later hominids. The brain is also extremely small, 410 cubic centimeters, one of the smallest of all known hominids. The face, however, is vintage *boisei*, large and dish-shaped. And, of course, there are those enormous cheek teeth. All of these *boisei*-like features are highly derived; that is, they have undergone a considerable amount of evolution from the basic ancestral pattern. They are the end product of the robust australopithecine lineage, not present at its beginning.

The unexpected combination of features in the Black Skull, then, is a mix of the very primitive with the very derived. The question was what we should call it. What species did the Black Skull belong to?

When I returned to Nariokotome for the final visit of the season, on September 13, Alan and I knew we had to confront the question. By this time work on the excavation was winding down. "There are two options," said Alan when we could avoid the subject no longer. We were sitting after dark at a table in the mess tent. The Black Skull looked even more bronze as it rested on the table between us, reflecting the harsh light of the hissing pressure lamp. "Either we call it *boisei*, in which case everyone is going to call us cowards, or we say it's a new species, in which case there's already a name out there for it." By this time our initial enthusiasm for the idea of a new species had ebbed considerably; we hadn't been completely serious about *Australopithecus kamoyensis*, but it did reveal our inclinations in those heady few weeks. "I think we should be conservative," I replied. "We'll be criticized whatever we do."

We knew that if we said the Black Skull was *Australopithecus boisei*, we would be stretching the idea of what *boisei* was like. True, the Black Skull had a classic *boisei* face in many respects, but it jutted out more than in later individuals, and it had those odd orbits. There was also the primitive brain case. We faced the classic problem of trying to chop up a lineage through time. As Alan said, "If you have a process of gradual evolution going on, starting with something like the Black Skull and finishing with the later *boisei*, where do you draw the line and say one species turns into another?"

Evolutionary biologists have wrestled with this problem for a long time, without a satisfactory solution. Chronospecies is the word used for different segments of an evolving lineage that is arbitrarily divided in time. I don't care for the notion, because it obscures the interesting changes in biology, the gradual evolution of parts of the anatomy. And the changes take place for a reason; they are a response to the pressures of natural selection.

In the case of the Black Skull–*boisei* lineage, the changes through time included an increase in the size of the brain and the shape of the back of the brain case, a face that jutted out less and less, and a more flexed, or bent, base of the cranium. Probably most of these changes were part of a single functional package, many different details of anatomy that evolved in concert. The most likely explanation was a refinement of the chewing

129

apparatus. When the face was shortened, the teeth were brought closer to the jaw joint, which increased chewing efficiency.

This same suite of changes also occurred in the *africanus-robustus* lineage and in the *Homo habilis–erectus* lineage. In each case the chewing efficiency was increased by a shortening of the face. Fred Grine of the State University of New York said, "This is the most incredible functional convergence I've ever seen." When you see the same evolutionary package unfolding independently in three separate lineages, you know something extremely powerful was going on. Exactly what it was, we can't say, because in the two australopithecine lineages the changes were accompanied by an increase in the size of the cheek teeth; in the *Homo* lineage, the cheek teeth got smaller.

When Frank Brown finally pinned down the age of the fossil, he came up with a very precise figure: 2.50 +/– 0.07 million years. It just so happens that 2.5 million fits right into a period that a group of geologists and paleontologists have identified as one experiencing a major climatic change, a global cooling. The modern form of Antarctica was established then, and the Arctic ice cap developed. On its own, this might not mean much. But as Elisabeth Vrba of Yale University has shown, 2.5 million years ago is also the time of great evolutionary activity among African antelopes, a group whose extensive fossil record she has studied in detail.

"Climatic change, through changing vegetation cover and population distribution, can drive evolution," she says. Could it be that the Black Skull–*boisei* lineage owes its origin to the same kind of environmental influences that produced a burst of new species among the antelopes? And there is tantalizing fossil evidence that the *Homo* lineage had its origin here too. Speculation, to be sure, but these are the kinds of phenomena that interest me in human prehistory, these insights into the biology of our species, not so much the names we give to things. Unfortunately, name things we must.

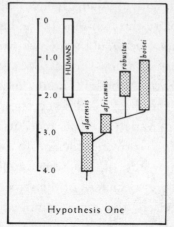

Hypothesis One

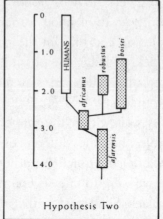

Hypothesis Two

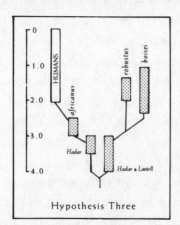

Hypothesis Three

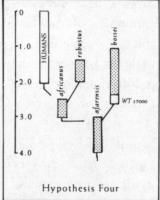

Hypothesis Four

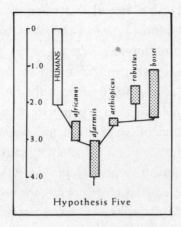

Hypothesis Five

"Yes, let's go with *boisei*," I said to Alan. "It gives one the flexibility to think about variability through time in the lineage." We also agreed that, in the paper we would write, we would at least acknowledge the possibility of a new species. We carefully crafted the following statement: "Although future finds may show that KNM–WT 17000 [the Black Skull's accession number] is well within the range of variation of *A. boisei*, it is also possible that the differences will prove sufficient to warrant specific distinction." In other words, if, by some good fortune, we were to find several more specimens of the same age as the Black Skull and with precisely the same combination of characters, then we would have a strong argument for saying there was a new, pre-*boisei* species here. But if hyper-robust contemporaries of the Black Skull turned out to be variable anatomically—some like classic *boisei*, some like the Black Skull, and some in between—then the argument for chopping the lineage into different species would be weak. We decided it was wisest to wait and see. "Fred Grine won't like it much," predicted Alan.

He didn't. "Based on its differences from other known specimens, if this isn't a new species, we're in serious trouble as far as identifying *any* new species," he told a journalist for *Science News*. Fred wasn't alone in urging us to be bolder. Six months after the end of the season, Alan attended a scientific gathering in San Francisco and took along a cast of the Black Skull. "No surprise, it generated a lot of interest," recalls Alan. "But people kept warning me, 'If you don't give this a new species name, someone else will.'" This was in April, four months before our conservative position was published in a paper in *Nature*.

Sure enough, someone else did name a new species: *Australopithecus walkeri*, "in honor of its discoverer," said the author of the name, Walter Ferguson, a technician at Tel Aviv University. "What a joke," scoffed Alan. "Ferguson's always doing things like that. If the Black Skull really is a new species, then it has to be

132

Australopithecus aethiopicus." We had said as much in our *Nature* paper.

The species name *aethiopicus* has been around for more than two decades, and for me there is an interesting irony. It goes back to the 1967 joint Kenyan-French-American expedition to the Lower Omo Valley in Ethiopia, the one from which I pulled out and went on to launch the Koobi Fora project. During the Omo expedition, the French team found a battered, completely toothless lower jaw of a hominid. Not much on which to base a new species, but its discoverers had little choice, because there wasn't much else comparable to it at that age, about 2.6 million years. Camille Arambourg and Yves Coppens named it *aethiopicus*, attached to the genus name *Paraustralopithecus* (near-*Australopithecus*), because of the uncertainty at the time as to whether it was truly *Australopithecus.*

This little jaw is a good example of the problems of trying to identify species unequivocally in the fossil record, particularly when the specimens are fragmentary. For instance, some authorities, including Johanson and White, have since argued that, no, it is not a new species at all; it is simply an Ethiopian *africanus.* Other equally renowned authorities have said, no, that's not right either; it is *boisei.* The little French jaw has therefore suffered something of an identity crisis in recent years. But, through the Byzantine rules of zoological nomenclature, the Black Skull may be the savior of *aethiopicus.*

The jaw is from a much smaller individual than the Black Skull, so it is difficult to compare the two directly. In any case, apart from overall shape and size, and the size of the teeth, there is no direct comparison to be made. Period. But, as luck would have it, a couple of days before the end of the Black Skull season, Mwongella found a hominid lower jaw at Kangatukuseo, just two miles southeast of the Black Skull site. The new jaw is marginally younger than the Black Skull and is massively built, just as one would expect to find in a Black Skull type of individual. The key

133

point, however, is that although the jaw is from a bigger individual than the *aethiopicus* specimen, and is slightly narrower at the front, the anatomical pattern is essentially the same. This, it seemed to Alan and me, would provide a link between the Black Skull and an existing species name, *aethiopicus*. And the rules say that if the anatomy of a new find matches that of an existing fossil, you must use an existing species name. So if the Black Skull is not *Australopithecus boisei*, it has to be *Australopithecus aethiopicus*.

When we made our discovery public—there was no leak this time—pictures of the Black Skull appeared in newspapers throughout the world, some with Alan alongside in a pinstriped suit, a rare sight. And it did, as a reporter for the *New York Times* put it, "shake some old branches on the family tree of early man." In our *Nature* paper, we said the same thing, but rather more formally: "Whatever the final answer, these new specimens suggest that early hominid phylogeny has not yet been finally established and that it will prove to be more complex than has been stated."

What we meant was that the pattern of our evolutionary history in the period three to two million years ago had sprouted another branch. To the three branches we mentioned earlier—*Homo habilis*, *Australopithecus robustus*, and the 1813–OH 13 type—there was added a fourth, the Black Skull–*boisei* type. Our family tree was getting bushier and bushier, and the west Turkana expeditions had played a major role in its growth. I was very satisfied with that.

But, no, we never did find the Turkana boy's hands and feet.

∎

In

Search

of

Humanity

∎

8

Human Origins

I have always thought it reasonable to imagine early hominid social life as analogous, in some strictly circumscribed ways, to the social life of savannah baboons. Baboons live in troops, some small, some comprising as many as a hundred individuals. My guess is that, because early hominids were bigger than baboons, their troop size was smaller. Nevertheless, their habitat would have been similar: patchy, open woodland and some gallery forest, offering a range of plant foods: nuts, fruits, shoots. No doubt they foraged for grubs and birds' eggs, just as baboons do today. No doubt, too, they occasionally captured young antelope and the young of monkeys, just as baboons do today. And I can see no biological reason why, if a troop of *Australopithecus* adults came across an infant of another australopithecine species or even an infant *Homo*, they wouldn't dispatch and eat it. There is no evidence, though, that meat was a significant component in the diet of early hominids or in australopithecines in general.

Mornings would have started for australopithecines as they do for baboons. The australopithecine troop wakes as the sun rises, individuals in their sleeping places in trees or along cliff-face perches, relatively safe from night-time predators. Slowly they clamber to the ground to the accompaniment of scattered vocalizations and occasional skirmishes, the unfinished business of the previous day's social interactions. The few mature males, unrelated to one another, wary as usual, forage desultorily and gaze around, taking stock of the new day. These males, stockily built and muscular, are much bigger than the females in the troop, the difference being about twofold. One of the mature males, who joined the troop only recently, hangs around at the edge of things, not yet fully accepted by troop members, especially by the mature males.

The mature females—sisters, half-sisters, and more distant relations—outnumber the mature males. As always in the early morning, this band of more or less related mothers is busily inspecting their offspring, the younger of whom are already beginning the serious business of play: running, chasing, fighting, sometimes screaming in fear as antics become threatening. One of the females has recently gone into estrus and is ready to conceive. Alerted by the signals, one of the big males has been following her about for days, occasionally copulating with her, always ready to chase away other males who try to sneak a copulation. There is a reason the males are so much bigger than the females: they compete with each other for sexual access to females, and physical strength is important. The females are aware of this and favor the bigger males. It is part of their biology, part of their evolutionary milieu.

When the sleepiness of the night has passed and the morning routine is completed, the troop begins to move off in the direction of a patch of trees that, about this time of year, is coming into fruit. Alert to the threat from neighboring troops, our band makes a leisurely path to its destination for the day. A water source is nearby, and trees for shelter and rest at midday. By day's end the troop will have covered seven miles or so, not in a straight line, but as a wandering forage, visiting known sources of food, checking others to see how long before fruiting will occur. As night falls the troop settles down to rest, this time in a different set of sleeping trees, one of many that is visited repeatedly throughout the course of a year. Occasional noisy quarrels eventually subside, and sleep comes.

Several neighboring australopithecine troops had been spotted during the day, provoking alarm calls between them. Some of the young males take a special interest in their neighbors, because when they near maturity they will have to seek a new home: the big males with whom they grew up will kick them out. The young males therefore see the neighboring troops not just as a source of potential aggression but also as a potential future home. It will be a difficult time for them, the eventual transfer.

In this sketch the principal difference between these hominids' behavior and what we see in modern savannah baboons is the hominids' mode of locomotion: bipedal as against quadrupedal. Everything else is imagined to be similar to any large primate that forages in relatively open country for a largely vegetarian diet. Here, remember, we are describing the life of the earliest hominids and all later australopithecines. How might it have changed with the arrival of the genus *Homo?*

From a distance, a troop of *Homo habilis* or early *Homo erectus* looks much like a troop of any of the australopithecine species. Some twenty or thirty individuals, males and females, young and sexually mature. But from a closer vantage point, their larger brain case would be apparent, as would be the less protruding face, and a more slender, humanlike build. Most significant in the context of social structure and behavior, however, is the body-size dimorphism. The males do not tower over the females, as they do in australopithecine species: the difference in overall

139

size is now only about 20 percent. There are more mature males in the *Homo* troop, too, in relation to mature females. More subtle, but part of the same biological feature, there is less tension among the mature males, less confrontation, more friendly interaction, more cooperation. The underlying reason for all this is that the males are related to one another: brothers, half-brothers, and so on. They come from common genetic stock and have good biological reasons for working toward common ends.

We may find our early *Homo* troop waking to a new day aloft in sleeping trees or cliff ledges, as with the australopithecines. Or perhaps on terra firma, protected by a crude shelter, a ring of large stones supporting a barrier of branches. On average, the troop size is smaller than in australopithecines, yet another biological signal that something significant has changed with the evolution of *Homo*. That signal reverberates in the social setting and in the mode of subsistence.

Like all large primates, australopithecines and *Homo* would have been intensely social creatures. Bonds between kin are strong, particularly between mothers and offspring and among siblings. But strong friendships and alliances form too, sometimes as male-female cohorts, but more often as "political" alliances in the constant struggle for power and status. When *Homo* males reach maturity, they don't transfer to another troop, as baboons do and as we hypothesized was true of australopithecines; instead, they stay in their natal troop, brothers and cousins living together, cooperating with one another.

Another significant difference is that the social structure is more tightly knit, not least because the infants are more helpless and require more care and protection than australopithecine offspring. Also part of that richer fabric is a prolonged childhood: there seems much more to learn. And, yes, the term "vocalization" doesn't do justice to what we hear. Something akin to verbalization is more appropriate. The waking up process is therefore noisier than among australopithecines, and more complex.

Instead of a dilatory drift of troop members toward the day's goal, here we see a knot of mature females and some of their livelier offspring setting off in one direction, three or four mature males, unencumbered by youngsters, setting off in another. A small group is left behind, including some of the younger infants. The females are noisy as they go, the males more attentive, one of them pointing to vultures circling overhead some miles distant. The females are confident of returning laden with nutritious roots and fruits. If they are lucky, the males may bring back a bonanza of meat, which will be noisily shared by all members of the band.

I could continue with this scenario, but it is already clear that with *Homo* we are looking at a putatively very different kind of animal from *Australopithecus*. The direction of that difference is the evolutionary springboard on which we once stood, to use an analogy illuminated by hindsight. Bipedal apes had been in existence for a long time when *Homo* arrived. The human family emerged about 7.5 million years ago; *Homo* evolved sometime before two million years ago. There was a large evolutionary gap between the origin of the first hominid species and the origin of *Homo*, a gap as great as five million years.

It is difficult for us to look back at a five-million-year slice of evolutionary time with any true comprehension. It is especially difficult when that period is populated by creatures endowed with human characteristics. But the link that binds us with these earliest hominids, the association we so strongly feel, is their bipedalism: standing upright as we do makes them seem like us. It is true that modern humans are bipedal apes in a sense, but the earliest hominids were bipedal apes, and no more. Only with *Homo* did the evolutionary equation change, and in a dramatic direction. The earliest *Homo* were incipient hunters and gatherers, and that way of life would shape the human body and the human mind for more than two million years.

But how can we know the equation changed with *Homo*? What variable are we looking for? More particularly, we would like to have a sense of when and where a degree of humanness began to infuse our ancestors. Was it present from the beginning, swelling ever more forcefully? Or did it arise *de novo*, late in human history? What can we say about why it happened? And what can we say about the shape of the evolutionary bush, of which we are the sole surviving branch?

I believe we are beginning to find some answers. In a surprising and gratifying way, the Turkana boy provided the clues. The story begins with his teeth.

"It was fortunate for me that the Turkana boy died the way he did," says Holly Smith, an anthropologist at the University of Michigan and a specialist on fossil hominid teeth. Remember, the boy's body seems to have floated in shallow water for some weeks after he died, during which time the soft tissues putrefied and all the teeth fell out of the jaw. Alan was therefore able to make exquisite casts of the teeth before he put them back into the jaw and assembled the skeleton. "They are beautiful casts," Holly says. "I was able to get all the information I needed from the roots as well as the crowns." I had asked Holly whether she would be responsible for studying the Turkana boy's teeth, as part of a growing team of investigators who were helping in the analysis of this most remarkable evolutionary treasure. I was as thrilled as Holly was about the excellent quality of the information she would be able to extract from Alan's casts. However, I could not have predicted how very informative it would prove to be.

We all remember from our childhood the business of losing milk teeth—with the attendant gifts from the tooth fairy—and the pride of the emerging permanent teeth. But unless there is a special reason for us to be interested in the details of dentition, few of us realize that there is a particular pattern for the eruption of permanent teeth: first molar, first incisor, second incisor, first premolar, canine, second premolar, second molar, third molar. There is some variation between individuals, of course, but overall the pattern is crisp.

For anthropologists, at least as important is that this pattern follows a clock, the main markers of which are the eruption of the molars: first molar at about six years of age, second molar at about eleven to twelve, and third molar at eighteen to twenty. So if I am shown a human jaw in which the second molar has just erupted, I know the individual died when he or she was about eleven or twelve. In the Turkana boy's jaw, the second molar was just in the process of erupting, which is why our initial estimate of his age was twelve years.

"You can get a more reliable estimate of age if you're able to look at the stage of development of roots and crowns in various teeth," Holly said at the time. "It's best if you don't have to rely on just the eruption of the molars." Hence the good fortune that the boy had shed his teeth shortly after he died, and the even better fortune that we managed to find them all. Holly was able to get information on all parts of the teeth. So how old was the boy when he died? Given that we have so much dental information to go on, the answer surely must be straightforward and unequivocal. "Well," equivocated Holly, "it depends."

The reason for the uncertainty rested squarely on the question we were trying to answer: In a biological sense, how human was *Homo erectus* 1.6 million years ago? "If the Turkana boy followed completely the human trajectory of development, then he was about eleven when he died," explained Holly. This somewhat younger age than our original estimate was based on detailed information from all the teeth, not just the eruption of the second molar. "And because we know that human adolescents go through a growth spurt—between thirteen and fifteen for boys —we can predict that the boy would have had considerable growth remaining, some 23 percent of his present size, according to human standards. He was 160 centimeters when he died, so that would have put him on track for being 198 centimeters as an adult, or about six feet four inches."

But maybe *Homo erectus* did not follow the human trajectory of development. Perhaps his development was more like an ape's. The difference between human and ape development is significant, both in degree and in what it implies for sociality.

Barry Bogin, also at the University of Michigan, has made a special study of human growth in comparison with other mammals, particularly other primates. He explains, "The pattern of human growth is characterized by a prolonged period of infant dependency, an extended period of childhood and juvenile growth, and a rapid acceleration in growth velocity at adolescence that leads to physical and sexual maturation. This pattern is unusual for mammals in that most mammalian species progress from the infant nursing period to adulthood without any intervening stages."

Primates are a little different from the general mammalian pattern, because the childhood period is extended. But there is no adolescent growth spurt in nonhuman primates. That is a human characteristic, and any parent whose offspring have gone through this stage knows how dramatic it is. One minute the child is just that—a child—and the next, he or she is an adult, once the gangling awkwardness is over. If you happen to be a chimp parent, you won't see this overnight transformation in your offspring. Instead, you would see a more steady transition. Why is human adolescence so special?

"The extremely extended childhood in humans is the result of a much reduced rate of growth during that period," explains Bogin. "The brain, however, is an exception during this period. It achieves virtually adult size when body growth is only 40 percent complete." This, he suggests, is the key. "These growth patterns help establish teacher-student roles that remain stable for a decade or more, allowing a great deal of learning, practice, and modification of survival skills to occur." Humans become human through intense learning—not just of survival in the practical world, but of customs and social mores, kinship and social laws. In other words: culture. Culture can be said to be *the* human adaptation.

In traditional societies, childhood is the time when much of culture is absorbed, often through rites of passage. Bogin suggests that the efficiency of the teaching process is enhanced if the offspring remains childlike for as long as possible; hence the extended childhood. Once this period is at an end, there is a lot of catching up to do in terms of physical growth. This is what's accomplished during the adolescent growth spurt, a brief burst of rapid growth that puts the development trajectory back on track. A human adolescent who is about to embark on its growth spurt can expect to increase in size by almost a quarter. By comparison, an adolescent chimpanzee at an equivalent stage in its life will grow by only about 14 percent to adulthood, because there is no growth spurt.

"If early *Homo erectus* followed the chimpanzee trajectory of development, not the human one, then our analysis of the Turkana boy would be different," Holly says. "For a start, development proceeds much faster in chimpanzees, so the second molar eruption is closer to seven years than to eleven years." What does this imply for the Turkana boy, given the ape trajectory of growth? It would mean that he was only about seven years old when he died, and that he only had another 14 percent to add to his growth. His projected adult height would have been 182 centimeters, or six feet. "This is smaller than on the human trajectory," comments Holly, "but it's still tall by any standards."

I was delighted to see Holly bringing these precise biological considerations to human fossils. Too often we anthropologists have simply made assumptions about the apeness or the human-ness of early human species. I was guilty of doing that, of course, when I guessed that the Turkana boy was twelve when he died: I had applied the human model of development to his stature and likely growth. But Holly insisted that we should be able to re-place assumptions by deductions. If she was right, then we really could be more definite about the emergence of elements of hu-manity during our evolutionary history, not just the emergence of elements of human biology.

What we needed to know, then, was whether the growth trajectory in early *Homo erectus* followed the human or the ape pattern, or maybe something in between. We would feel on safe analytical ground if we could say that *Homo erectus* was "just like humans" or "just like chimpanzees," because we have precise models in front of us: we can measure what humans do or what chimps do, and then deduce what early *Homo erectus* did. But just as we've learned that the anatomy of extinct species is not a precise copy of various forms of modern anatomy, we are likely to find that physiology isn't either. If the Turkana boy was like neither modern humans nor modern apes in his growth pattern, how can we find out what he was like? Holly Smith may have provided the answer for us in a most unexpected way.

In September 1986 she published a paper in *Nature* on pat-terns of tooth development in hominids. The paper ignited a lively debate among anthropologists, the public part of which was played out on the pages of both *Nature* and *Science*. Holly sent a manuscript copy of her paper to David Pilbeam, who, she thought, would be interested in it. He was, and he suggested that she be invited to take part in a symposium, "Behavior of the Earliest Hominids," to be held in Sweden in March 1988.

"My *Nature* paper was on dental development, not behavior, so I decided I'd better start thinking what it might mean in terms of behavior," Holly now recalls. "I went to the library, got out a big paper by Paul Harvey and Tim Clutton-Brock, and started reading about life histories." She also came across a paper titled "The Evolution of the Human Brain," which, four years earlier, the British anthropologist Bob Martin had delivered as a major lecture at the American Museum of Natural History. In it, Bob analyzed brain evolution not so much in terms of "increased intelligence," as has often been done, but in the context of basic biology, of how a species is metabolically and ecologically able to build a brain of a particular size. Martin's analysis was very much in the spirit of Harvey and Clutton-Brock's approach. "Bob Martin's paper was important," says Holly. "It influenced me a lot." Several lines of investigation were beginning to come together here, and Holly was poised to exploit them brilliantly. The key was the notion of life history evolution.

I've used the phrase "life history" several times but have never fully explained it. Essentially, it is a description of how an animal lives, not what it eats or what it spends its days doing, but the pattern of its life and death. Among the factors of life history are how quickly offspring are weaned, the age at sexual maturity, gestation length, the number of offspring per litter, the time between litters, and longevity. The study of how these factors relate to one another in individual species has become a hot area in behavioral biology research.

Harvey and Clutton-Brock, British evolutionary ecologists, opened the paper to which Holly referred like this: "G. Evelyn Hutchinson once argued that priorities for ecological research should include the questions 'How big is it and how fast does it happen?'" These questions apply at all levels of life history, but they also point to the pattern. In other words, the bigger a species is, the slower things happen: longer gestation, longer time to weaning, longer time to maturity, greater longevity. Roughly speaking, large species live slow lives; small species live fast lives.

For instance, the gorilla is a large species in the mammalian

147

world; its life history is slow. A female will have her first pregnancy when she is about ten years old and can expect to live to the age of forty. At the other end of the scale is the mouse lemur, the smallest of all primates, which weighs in adulthood a featherlight 80 grams. Females produce their first offspring when they are nine and a half months old and have a life expectancy of fifteen years. A dramatic consequence of this great difference in speed of living is the production of descendants. Imagine a female gorilla and a female mouse lemur that were born on the same day. The mouse lemur would have matured, had its offspring, and died, leaving descendants who would multiply to ten million individuals, before the gorilla would have had its first offspring. Nevertheless, the hearts of both animals would, at the end of their respective lives, have pumped approximately the same number of times.

A second aspect of life history biology is a species' potential reproductive output, or how many offspring an individual could theoretically produce in its lifetime. Some species produce many offspring, to which they devote little parental attention. The potential reproductive output of each female is high, but many offspring never make it to adulthood. Female salmon, for instance, produce millions of offspring (eggs that are fertilized) and therefore have an enormous potential lifetime reproductive output. However, most of the offspring finish up in the gut of one kind of predator or another. Other species, by contrast, produce few offspring and lavish a great deal of attention on them: gorillas, for example, in which the potential lifetime reproductive output is only about six. Here, although each offspring stands a much better chance of surviving to maturity, reproductive output for a mature female is never very high. A female salmon and a female gorilla, at the end of their respective lives, will have replaced themselves and one other: the strategies for achieving that end are, however, very different.

Ecologists find that high potential reproductive output is an adaptation to unstable, unpredictable environments. Species with low potential reproductive output, by contrast, are adapted to stable, predictable environments. Primates as a whole are toward the low potential reproductive output end of the spectrum, with apes and humans at the most extreme. Presumably, this tells us something about the ecological circumstances under which humans evolved; namely, they were fairly stable and predictable.

But with only fossils to go on, how can we get an insight into all this with respect to human ancestors? Can we ever know the longevity of early *Homo erectus,* for example, or whether the Turkana boy was weaned when he was about three years old, as a human infant would be? Holly was determined to find out. "I'm a tooth person. I've always been a dental anthropologist, so I wanted to see if I could express some of this life history theory in terms of teeth. It would mean something more to me then."

But it would work only if stages of tooth development accurately reflect life history patterns. "The dentition must be closely integrated into the overall plan of growth and development," Holly points out. "After all, the teeth process the food that fuels all growth. Teeth must emerge so that babies can be weaned, permanent ones must replace deciduous precursors before they wear out, and molars can't emerge before there's sufficient growth in the length of the face. For the survival of the individual, timing of dental development is critical." That was the reason for hoping that the development of dentition would indeed echo life history patterns.

Holly first immersed herself in the Harvey and Clutton-Brock treatise, which contained longevity, age at sexual maturity, gestation length—all the available life history data for as many primates as had been studied. She was then ready to test her conviction that teeth could provide a way into it for those of us who deal with fossils. She wanted to see how well the age of eruption of the first molar correlated with life history factors. First molars were promising, because they are the earliest permanent teeth to erupt in primates and in many other mammals, too, and they are stable in many aspects of their growth. "I dug into the literature and found data for first molar eruption in twenty-one primate species, and I plugged this into my computer, together with the relevant data from Harvey and Clutton-Brock. I just sat there and looked at the results. It was quite surprising."

Holly was looking at this string of numbers: 0.89, 0.85, 0.93, 0.82, 0.86, 0.85. These are correlation coefficients between the age of eruption of the first molar and the life history factors of body weight, length of gestation, age of weaning, birth interval, sexual maturity, and life span. "They were *very* good," she comments with pleasure. A high correlation coefficient means that the two things being compared are closely tied to each other. The highest correlation coefficient you can get is 1.00, which means that the two things you are looking at are marching together in lockstep. Holly's figures made it clear that dentition—something we can see in the fossil record—really does give an insight into, say, length of gestation, age of weaning, and so on. But the tightest correlation coefficient of all was with the size of the brain: 0.98. "That was fantastic."

Brain size is turning out to be something of a touchstone in life history theory. Some years ago, George Sacher, a zoologist at Argonne National Laboratory in Illinois, said that the size of a species' brain determines very closely the species' growth patterns. Harvey and Clutton-Brock's analyses expanded this key insight, showing that brain weight is a good predictor. If you know a species' brain weight, you can, with the appropriate mathematics, calculate each of its life history variables. You can know its longevity, its age of weaning, the interbirth interval, and more. And now Holly's work has effectively tied all this into the fossil record via those reliable teeth. It gives us a way of peering into the past, a way of seeing real creatures, real biology, not just petrified bones.

I found it thrilling to watch Holly weave together these disparate lines of argument. In one realm she was working with elements of our past with which I am comfortably familiar, the fossils; in another, she was drawing on research in which I am only a spectator, life history theory. I could see that the approach promised to teach us things about human evolution that once were unreachable.

"The approach becomes powerful for our purposes, because we can get good estimates of brain size and, to a lesser degree, body size, from the fossil record," comments Holly. "I looked at brain and body weight data for a range of fossil hominids, and got predictions for age of first molar eruption and life span. You see a very interesting pattern." The pattern included three grades of hominid. The first is what Holly calls a "chimpanzee grade of life history," which applies to all australopithecines. In the third grade, life history approaches and then achieves that of modern humans. Later *Homo erectus*, Neanderthals, and, of course, *Homo sapiens* are in this group. Between the ape pattern on one side and the human pattern on the other is what Holly describes as an "intermediate life history." It is here we find the first known *Homo* species, *Homo habilis*, and early *Homo erectus*, which includes the Turkana boy.

Compared with modern humans, in which the first molar eruption takes place at 5.9 years and life span is sixty-six years, early *erectus* has figures of 4.6 years for first molar and fifty-two years for life span. For the australopithecines, the figures are a little over three years for first molar eruption, and a span of about forty, just as it is for chimpanzees. As Holly notes, "Early *Homo erectus* is truly in an intermediate position." The species was neither "like a human" nor "like an ape"; it was a species with its own life history pattern. Using this information, we can calculate that the Turkana boy actually died when he was nine years old, not eleven (the prediction from the human pattern) or seven (the prediction from the chimpanzee pattern). He would have weaned at something less than four years, and would have become sexually mature at about fourteen to fifteen years. His mother probably had her first baby when she was thirteen or fourteen, after a nine-month pregnancy. After that, she probably was pregnant every three or four years.

It is remarkable that we can discuss the biology of an extinct human species with this degree of detail. If our extrapolations are correct, we have to reshape our perspectives, particularly on the early hominids, the australopithecines. For a long time, people talked about the origin of the human family in terms of human-like characteristics—hunting, technology, social life. In other words, the assumption that all hominids were to some degree recognizably human has pervaded anthropology. Even when these specific explanations are replaced by more biologically based hypotheses, the notion that hominid—*all* hominids—always were "like us" in some way persists still, albeit less explicitly. In my sketch, I deliberately made the australopithecines bipedal apes, and no more; not miniature humans. Yet early *Homo* was beginning to look a bit like us, at least in its biology. I think that division is probably correct. We can begin to think about the emerging human social milieu in early *Homo*, the milieu in which our humanness surely began to be formed.

But how confident could we be that the calculations from life history theory were valid for extinct hominids? How confident could we be that australopithecines were bipedal apes; that *Homo* was a humanized bipedal ape? The analysis so far, impressive though it was, provided only one line of evidence. Independent confirmation would help to build confidence. As it turned out, at least two such lines of evidence were to back it up; and once again teeth would open a revealing window onto the past.

9

This Way Lies
Humanity

O n l y   r a r e l y does a piece of research influence a
field deeply and for a long time. Alan Mann's analysis of hominid
dental growth patterns, which he carried out in the early 1970s,
is one such example. "His work established conventional thinking for two decades," observes Christopher Dean, an anthropologist at University College, London. "What he said was important, because it affected the way we thought about early
hominids, about how human we thought they were."

Mann, an anthropologist at the University of Pennsylvania,
had examined juvenile hominid jaws from the cave site of Swart-
krans, on the Transvaal in South Africa, looking at the pattern of
development and tooth eruption. He concluded that the pattern
was virtually the same as in modern humans, and not at all apelike. "It tremendously reinforced the already strong belief that all
hominids were humanlike in many ways, even the earliest
hominids," says Dean. Models for all early hominids were scaled-
down versions of modern hunter-gatherers.

When, in October 1985, Chris Dean published a paper in *Nature* saying that Alan Mann was wrong, it was therefore a challenge to prevailing paleoanthropological wisdom, not just to one person's opinion.

A challenge to conventional wisdom in science, no matter what field, is no trivial matter; it usually invites vigorous and sometimes scathing rebuke. This was to be no exception. Occasionally the challenge withstands the attacks, and, by steadily gathering support from other directions, eventually replaces the accepted view. It was into this process of replacement that the Turkana boy's skeleton was thrust. He is part of what, in my view, will one day become the new conventional wisdom, which encompasses some of the insights we have glimpsed.

The process of replacement was as complex as it was unpredicted, and involved journeys through aspects of our ancestors' anatomy that at first sight seem unconnected but turned out in wonderful ways to be joined by a common road. That road led eventually to the newly emerged hominid adaptation, the beginnings of a hunting-and-gathering subsistence, and linked tooth anatomy, facial architecture, enlarged brains, and tool making in a harmonious whole. If there were a signpost on that road, it would read: THIS WAY LIES HUMANITY.

We have seen how Holly Smith's discovery of three grades of life history patterns in hominids indicated that conventional anthropological wisdom might be wrong. There was the human grade, which we see in modern humans, Neanderthals, and later *Homo erectus*. There was the ape grade, which all australopithecines display. And there was an intermediate grade, which occurs in early *Homo*. Holly's discovery indicated that not all hominids followed the human grade, and also that we could identify the point in our history when a shift from ape to human grade took place. It was a discovery of major proportions in anthropology, and it required corroboration. Chris Dean's work seemed to offer

the kind of support we sought.

Dean had coauthored his 1985 *Nature* paper with his colleague Tim Bromage, an anthropologist at Hunter College in New York. The main thrust of the paper—which was called "Reevaluation of the Age at Death of Immature Fossil Hominids"—was that tooth development did not follow a single pattern, the human pattern, as claimed by Alan Mann. Instead, the human pattern of development is to be found in modern humans, Neanderthals, and late *erectus*; in the australopithecines, the ape pattern prevailed; in early *erectus*, something intermediate occurred. Note how this neatly echoes Holly's discovery on life history patterns. Dean and Bromage's discovery on the pattern of tooth development clearly was going to influence the rising debate about the place of humanness in human prehistory.

The two young researchers needed to ascertain how old a particular fossil individual was when it died. Teeth provided the evidence. Measuring age by using teeth is a bit like measuring the age of trees: count lines. Under a microscope the enamel surface of the tip of an incisor tooth is seen to be regularly rippled, the ripples separated from one another by lines—perikymata, as they are known. Although there is a great deal of uncertainty about the nature of the lines, it is thought that a ripple—the space between one line and the next—represents about seven days' growth. If you count the lines, and make various necessary adjustments, you can calculate age.

The most obvious specimen to begin with was the Taung child, which was said to have been about seven years old when it died, based on a human pattern of dental development. "As it turned out, it wasn't possible to do this aging on the Taung specimen," explains Bromage, "but we were able to do it on another jaw that was at the same stage of dental development. The answer we got was 3.3 years, which is exactly what you'd expect for an ape whose first molar is just erupting. It most definitely was not what you'd expect in a human child with the first molar just emerging."

Once they had got the completely unexpected result for the Taung child's age, Bromage and Dean decided to apply the technique to other hominid specimens, including early *Homo*. The message was clear. All the australopithecines developed their dentitions quickly, like apes. Even early *Homo* was not significantly slowed down, showing that prolongation of the childhood growth period was not marked. "These results reflect a dramatic advance in our appreciation of the biology of these species," wrote Bromage and Dean in their paper. In other words, we should be thinking "ape biology" not "human biology" when we try to understand our earliest ancestors.

Bromage had become involved in this tooth work as a side issue to his own interest in analyzing the growth pattern of early hominid faces. Humans and apes construct the architecture of the face in different ways during early life, which gives another potential marker for judging apeness and humanness in the early hominids. Following the rising drum beat of other evidence, Bromage found that australopithecine faces are built like ape faces, not like human faces. The pattern changes only with the evolution of *Homo*.

So the life history data said it. The tooth development data said it. And the facial development data said it. The "it" was going to revolutionize anthropology: a significant shift in *biology* took place with the origin of *Homo*; it was not just a shift in *anatomy*. In other words, the evolution of *Homo* involved more than simply a change in the size or the shape of the body or a part of the hominid body. The pattern of growth, longevity, and so on underwent a change with the arrival of *Homo*, a change that introduced humanlike biology for the first time among our ancestors. Were there more lines of evidence to back this up?

Maybe so, and again the teeth could be a source of insight.

Bromage and Dean had been looking at the *rate* of dental development: in early hominids, teeth developed rapidly, as in apes; in *Homo*, the rate of development began to slow down. Holly Smith also decided to look at tooth development, but at the *pattern*, not the rate. By pattern she meant the order of tooth formation. Was it, in early hominids, like humans' or like apes'?

The most notable difference in the sequence of tooth appearance is that in apes, the canine erupts after the second molar; in humans, it precedes it. Moreover, humans have shifted the entire relation between the development of the anterior and posterior teeth. Holly's task was to see which pattern of development was closer to that seen in early hominid jaws.

"The really hard work was in establishing good standards, and I discovered that Chris Dean and Bernard Wood had done that in 1981," she says with relief. "I was really grateful to them. All I had to do was plug my data in and get out the results. It just seemed to fall into place." There were some complications, but the message, once again, was that early hominids developed like apes. In *Australopithecus afarensis* and *Australopithecus africanus*, the dental development pattern was clearly that of apes, and even *Homo habilis* was apelike. In early *Homo erectus*, however, a shift toward a humanlike pattern was already discernible.

"Absolutely and totally wrong" is how Alan Mann described Holly's analysis to a reporter for *Science*, thus signaling the beginning of an intellectual battle that still continues. Bromage and Dean were drawn in, too. "Their analysis is absolutely and completely wrong," Mann states. "It is true that when I published my original analysis we were thinking about the australopithecines as little humans. Things have changed dramatically, but I would term it a fad to say that the australopithecines were just like apes."

A series of letters was exchanged in the correspondence columns of *Nature*. Alan Mann reiterated his view that "the current data reveal a clearly humanlike pattern" of dental development, and Holly Smith repeated hers, that "a whole series of fossils, spanning millions of years, share a fairly typical apelike pattern of development." The same evidence leading to diametrically opposed interpretations, strongly expressed: something was wrong. Perhaps new participants could clarify the debate.

That is what happened when Glenn Conroy and Michael Vannier, of Washington University in St. Louis, became involved. They came armed with the high-tech tool called computerized tomography, the CAT scan. Usually employed in hospitals for producing three-dimensional X-ray pictures of patients, CAT scanning can be exploited in studying people long dead. Conroy and Vannier's chosen subject was the Taung child, who had been dead perhaps two million years.

"The CAT scan gives us an opportunity to look inside fossils in a way that was impossible previously," says Conroy, an anthropologist. "We thought we could test the technique by looking at the Taung skull and maybe help settle the dispute." Vannier, a pioneer in developing scanning techniques in clinical medicine, was enthusiastic about turning his skills to subjects dead for two million years.

"The dental development patterns we find are very much like those of a three- to four-year-old great ape," says Conroy, describing what he and his colleague saw in the diminutive fossil. In other words, this independent line of evidence indicated that the australopithecines were indeed more like apes than like humans in their dental development. "The evidence is beginning to become overwhelming that Smith, as well as Bromage and Dean, is correct. Early hominids do appear to have had a more apelike dental maturation period."

A reassessment of the biological characteristics of the earliest hominids was clearly warranted by the new lines of evidence. But in order to counter a possible overreaction to it, Conroy and Vannier issued a caveat. "These great ape features must be weighed against the undoubted hominidlike features stressed by Dart in his initial description of the skull," they wrote. These include a lack of the large browridges you find in apes, the shape of the eye sockets, the lack of a gap between the canine and premolar teeth, which is characteristic of apes, the overall shape of the brain, and the position of the foramen magnum, through which the spinal cord passes from the brain case into the vertebral column. "This complex mosaic of craniodental features shows that the Taung child is not a little human, but, just as important, it is not a little ape."

We therefore have to think of the earliest hominids as bipedal apes, with apelike life histories and apelike facial and dental developments. Only with the evolution of the genus *Homo*—which brought an increase in brain size and a reduction in the size of cheek teeth, a shortening of the face—did patterns begin to change. According to the pattern of dental development (Holly Smith's work) and the rate of dental development (Bromage and Dean's data), the period of childhood was beginning to be prolonged in *Homo*. Can we tell why, with the Turkana boy, for instance, a prolonged childhood was necessary? Were there other changes in early *Homo* biology that set off these changes in the teeth and face?

We know, as Barry Bogin emphasizes, that the extended childhood in modern humans has to do with intense and prolonged learning, the stuff of human culture. In contrast to other large primates, humans mature slowly and then reach adulthood in a burst of growth at the end of adolescence. But prolonged dependence is also a biological necessity, because human babies come into the world too early.

This may sound odd, but it has to do with the extraordinarily large brain of which we are so proud. The large brain is associated with a slowing down and lengthening of our species' life history factors, such as sexual maturity and longevity. If we were to do the calculations for length of gestation based on our brain size, we would arrive at a figure of close to twenty-one months, not nine months. As a result, for almost the first year after birth human babies live like embryos, growing very fast but remaining essentially helpless. It is easy to see the benefit of delayed maturity: it allows for learning through culture. But how can we account for the bizarre biological arrangement whereby we spend our first year of life in the relatively hazardous outside world rather than in the safety of the womb?

The size of the human pelvic opening—birth canal—has increased during human evolution, thus accommodating the expanded size of the brain. But there are engineering constraints that limit the size of the birth canal, constraints imposed by the demands of efficient bipedal locomotion. At some point, the expansion of the birth canal stopped, and expansion of neonatal brain size had to be accommodated by embryonic maturation outside the womb.

If humans were like apes in the way the brain grows, then a human neonate could expect at least to double its brain size by the time it was mature. Doubling of brain size from birth to maturity is standard in large primates. Given that the average adult human brain is about 1350 cubic centimeters, the baby's brain at birth would be 725 cc if we followed the typical primate pattern. But we don't. The average brain size at birth is in fact 385 cc, and even this strains the mechanics of the system often enough for birth to be much more hazardous in humans than it is in apes. As a result, the growth of the brain in human babies continues at a fast rate for about a year after birth, giving the effective gestation period of twenty-one months. In the end, the brain more than triples its size between birth and maturity.

Because human babies are forced to come into the world with a relatively unformed brain, they are much more helpless than ape babies. This alone effectively lengthens childhood, and demands a greater devotion to care taking in the social milieu.

The necessity for social learning then lengthens childhood still further. What can we say about early *Homo erectus?*

Bob Martin, who heads the Institute of Anthropology in Zurich, raised this question some years ago. "It seems likely that increasingly elaborate parental care was required in *Homo erectus* and then still more in *Homo sapiens* to cater for the increasingly helpless condition of the infant during the first months of postnatal life," he concluded. His reasoning went something like this.

Suppose that the size of the birth canal in early *Homo erectus* was the same as in modern humans and therefore had room for a newborn with a 385-cc brain. And suppose that a *Homo erectus* infant doubled its brain size to adulthood, as ape infants do. Would such a doubling produce a brain of the size we know *Homo erectus* possessed? If the answer is yes, then no lengthening of childhood would be necessary to allow for fast brain growth after birth. But the answer is no. If you slightly more than double the human neonate brain size of 385 cc, you come to about 800 cc, which is smaller than the average of about 900 cc for adult early *Homo erectus*. Brain growth would therefore have to continue at a high rate for a time after birth in order to achieve the extra brain capacity in the adult *Homo erectus*. Infant helplessness, and prolongation of childhood, would have already begun in early *Homo erectus*.

In his calculations Bob had deliberately stacked the argument against *Homo erectus* by assuming that its birth canal was as big as it is in modern humans. If the birth canal was in fact smaller, then the conclusion about extended childhood would be strengthened. Our discovery of the Turkana boy gave us a chance to check this, because his pelvis was so well preserved. The width of the pelvic opening in humans is similar in males and females. After reconstructing the boy's pelvis, Alan was able to measure the width of the pelvic opening: 10 centimeters, as compared with 12.5 cm in modern humans. The difference was more than enough to make Bob's tentative conclusion look very conservative indeed.

"If you make all kinds of assumptions and adjustments, you can estimate the average brain size for early *erectus* babies at birth," said Alan after working with these new numbers. "The figure you come up with is about 275 cc, which is not much more than half the modern average." More significant, this figure implied that *erectus* individuals had to triple the size of their brain between birth and maturity, giving a 900-cc brain in the adult. Tripling of brain size after birth is the human growth pattern, and must have been associated with an extended period of infant care. In a paper that Alan and I wrote about the discovery of the boy, we were able to conclude: "The initial guess that fast fetal growth rates continued after birth, producing increased dependency and prolonged childhood, seems to have been correct."

In other words, we are beginning to see the emergence of humanness—some real sense of "like us"—in the biology of *Homo*. Before the evolution of our own genus, hominids were human only in the way they stood and moved; they were bipedal. But biologically they were bipedal apes, not like us. I find it tremendously exciting when analyses like those of Holly Smith, Tim Bromage and Chris Dean, and Bob Martin produce patterns of past events. And I was thrilled when the information from the Turkana boy's skeleton was slotted into the equation, further crystallizing the pattern. From the boy's pelvis we were able to see that early *Homo erectus* babies were born "too early," which forced an extended childhood, leading to a slowdown in the rate of tooth development. But the linking of these apparently disparate aspects of our ancestors' biology goes even deeper.

When we described the types of hominids that lived between two and three million years ago, we saw that there were several coexisting species and that a major anatomical shift occurred. In the hominids before *Homo*, and in the australopithecines that for a while were contemporaries of our genus, the males were twice as big as the females. *Homo* males, though, were only 15 to 20 percent bigger. Extensive body-size dimorphism, like that in baboons, is usually associated with intense competition among the males for access to the females, and the males in the troop are usually genetically unrelated to one another. A reduction in body-size dimorphism, like that in chimpanzees, is usually associated with reduced competition among males for access to females, and the males are often genetically related.

There's no doubt in my mind that the change in body-size dimorphism we see between australopithecines and *Homo* indicates a significant shift in human history. Something important happened in the biology and behavior of our ancestors at this juncture. Can we bring together the reduction in body-size dimorphism and the change from an apelike to a humanlike life history pattern? I think we can.

Prominent in the mix is the enlarged brain, expanding in an evolutionary punctuation from close to 500 cc in australopithecines to more than 700 cc in early *Homo*. An almost 50 percent expansion in brain size in creatures of roughly the same body size is a biological signal about as dramatic as can be imagined. As significant to me is the concomitant shift in the life history. And, as our earlier sketch of australopithecine and *Homo* troop life implies, there was also an important change in subsistence. Here, the new constituent is meat, not as a rare item in the diet, but for the first time a substantial component. Is it a coincidence that we see stone tools enter the archeological record at about the same time as we judge *Homo* to have evolved, some 2.5 million or more years ago? I think not. I think we are seeing here the elements of an evolutionary package that in time led to *Homo sapiens*.

What made possible that appearance? Was it the demands of technology, which required more brain power? Was it the origin of spoken language, a greatly refined mode of communication?

Was it the exigencies of an intellectually exacting social nexus? Or was it something that none of us has yet identified? I suspect that the driving force was no single element, but rather a mélange, brought together as a new adaptation.

The initial appearance of the hominid family 7.5 million years ago coincided with global cooling and local geological events that fragmented and thinned the previously carpetlike forest cover in East Africa. I say "coincided with," but I am really implying a causal relationship: the origin of bipedal apedom was an adaptation to new, mosaic environments. What of the origin of the genus *Homo*? Does it "coincide" with anything significant? Yes, it does: another global cooling event, much bigger than before. Huge ice mountains built up in Antarctica close to 2.6 million years ago, and for the first time significant amounts of ice formed in the Arctic. The frigid grip produced cooler, drier climates in the rest of the globe, including the varied highland terrain of eastern Africa.

Such climatic changes break up habitats, and may drive pulses of extinction throughout the plant and animal worlds. But they may also cause speciation, the development of new species from isolated populations, adapting to new conditions. Among the African antelopes, whose fossil record is as good as any terrestrial vertebrates' can be, this pulse of extinction and speciation at around 2.6 million years ago is clearly seen. Suddenly, a range of existing species vanished and a crop of new ones appeared. Glimpses of this pattern are also seen, albeit less clearly, in other grazing and browsing animals of Africa. I suggest that it is to be found in hominids, with the evolution of the robust australopithecines and of *Homo*.

The fragmentation and drying of East African habitats that began earlier than 2.6 million years ago was a challenge and, in the strict Darwinian sense, an opportunity. A challenge, inasmuch as an animal population usually tries to remain with its favored habitat, even if it has to undertake long migrations. In this case, the species continues to exist. But sometimes a population may be unable to track its favored habitat, and it will die out. Only if all populations of a species die out does the species become extinct. The evolutionary "opportunity" arises when isolated populations persist under the new circumstances. New selection pressures may result in a new adaptation, and eventually a new species. Adaptations may move in several directions: the adaptation of hominids 2.6 million years ago, I think, went in at least two directions.

One was a further exaggeration of the basic hominid form. This resulted in the robust australopithecines, like the Black Skull individuals. These creatures were able to process large amounts of tough plant foods, the kind found in arid environments. The second direction was something of a breakthrough, one that is recognized by the appellation *Homo*. Because the traditional hominid diet became more difficult to subsist on, there was the potential for expanding the diet, not specializing it, as the robust australopithecines did. The expansion involved making meat an important food source, not just an occasional item, as it was with earlier hominids and is still for baboons and chimpanzees.

Although some anthropologists argue that regular meat eating was a late development in human history, I believe they are wrong. I see evidence for the expansion of the basic omnivorous hominid diet in the fossil record, in the archeological record, and, incidentally, in theoretical biology. In his analysis of human brain evolution Bob Martin points out what all good biologists know: the brain is an expensive item to maintain. It constitutes only 2 percent of the body bulk, yet consumes almost 20 percent of total energy. Bob extends the argument, saying that the brain not only is expensive to maintain, but is expensive to build. In other words, during the development of the embryo in the womb, a disproportionate amount of energy has to be diverted into brain building. Why should a mother bother to do this if she could produce twice as many offspring with smaller brains?

"Because brains are powerful organs," says Bob, knowingly stating the obvious. "As a result, a species will have as big a brain as it can afford." This seems a curious way of expressing it, like the choice of how luxurious a car one can buy. But it is meant in the context of the biology of life history, where each factor ties in closely with all the others. It is no accident of biology that, throughout evolutionary history, animals have become brainier: mammals are brainier than reptiles and amphibians. And within mammals, primates are the best endowed of all. Such an expensive piece of equipment as the brain would not show this overall evolutionary pattern if there were not significant benefits to be enjoyed from it. In humans, we see the trajectory spiraling off into new dimensions, eventually becoming the seat of self-awareness, through which each of us knows ourselves and our world.

The initial expansion of brain size in hominids, which established the genus *Homo*, was more mundane. It concerned an adaptation that required more complex behavior: the hunting-and-gathering way of life in embryo. But it also fueled itself, in a kind of positive feedback. Part of Bob Martin's thesis about a species' ability to afford a large brain is that it must have a stable environment, stable in terms of food supply. Stable and nutritionally rich. The robust australopithecines managed to stabilize their food supply in the new prevailing environment 2.5 million years ago, but their tough plant foods were not rich nutritionally. By broadening the diet to include meat, early *Homo* achieved both stability and rich nutrition. Meat represents high concentrations of calories, fat, and protein. This dietary shift in *Homo* drove the change in pattern of tooth development and facial shape. The links in the chain join up yet more closely.

Our ancestors achieved this dietary shift through technology, and thus opened the road to the potential—but, remember, not inevitable—development of yet bigger brains. Primates have great difficulty in getting at the meat of large, tough-skinned animals. With a sharp stone flake, however, even the toughest hide can be sliced through, literally opening up a new nutritional world. In a very real sense, by taking a crude hammer stone and striking it against a pebble to produce a small, sharp flake, our earliest *Homo* ancestors began to control their world in a way that no other creature had done before or has done since.

There is a spot on the western shore of Lake Turkana, about three miles north of where the Black Skull was found and five miles in from the present water's edge, that gives us a glimpse of the earliest stages of this breakthrough. A couple of years ago Peter Nzube, of Kamoya's team, found what he considered to be a potential archeological site. The French archeologist Hélène Roche describes the artifacts Peter found as "very crude cores," lumps of lava that look to the untrained eye as just that. Hélène saw that flakes had been knocked off, apparently deliberately. "You don't have a sense of the systematic removal of flakes," she explains. "Instead, you get the idea of clumsy, occasional flake making." The lava cobbles were excavated from a land surface a little less than 2.5 million years old, making these artifacts among the oldest known anywhere.

The site is known as Lokalelei, named after a nearby seasonal stream. Two and a half million years ago, there was also a stream there, a small tributary of the river that coursed through the Turkana Valley, where at the time there was no lake. The lava cores that Peter discovered had somehow found their way into that ancient stream, to be buried in the sandy bed for almost 2.5 million years. We can imagine that a small group of hominids came across the recently dead body of an antelope or other medium-sized grazer, perhaps led to it by vultures circling over-head. When the hominids arrived they were lucky to have the corpse to themselves; jackals and hyenas were just as alert to the promise of meat signaled by the vultures.

Quickly and clumsily striking a few flakes from cobbles they picked up as they neared the corpse, some of the group were ready for the meat quest. Others may have been occupied in keeping the smaller scavengers at bay for a short time, enough anyway to allow a limb or two of the corpse to be disarticulated with swift slashing motions of the flake, cutting through skin, meat, and tendons. Rewarded with chunks of meat and part of a limb or two, our group rapidly retreated, leaving other meat eaters to strip the skeleton clean. Soon the powerful jaws of hyenas reduced the bones to fragments. These master scavengers took their tools—their jaws and bone-crunching teeth—with them. The hominids left their tools behind, to become the beginnings of the archeological record.

That record marks the path of human history through a huge tract of time. It gives us a way of touching our past, a scatter of more or less crudely manufactured stone tools, the products of human hands. I am very much aware of this when I hold in my hand an ancient artifact, particularly one of the most ancient of all, from Lokalelei. Here, we have the fruits of ingenuity, part of the evolutionary package. Our ancestors made these tools, but, in a real sense, these tools made our ancestors. By the same token, they made us what we are today.

When I am asked, as I often am at the end of a public lecture, how I know who made the tools, I give an answer based on history and logic, not on direct evidence. It goes something like this. Stone artifacts appear in the record at about the same time as we judge the genus *Homo* to have evolved, somewhat earlier than 2.5 million years ago. This, I suggest, is no coincidence. It is also true that the robust australopithecines evolved at this time. Perhaps they too were tool makers. Indeed, one anthropologist, Randall Susman at the State University of New York at Stony Brook, argues that robust australopithecine hands had all the anatomical characteristics necessary for making tools. How can we know whether he is right? I'm not sure, and I don't think anyone can be.

We can be sure, however, that *Homo* was a tool maker, because once *Homo* species were the only hominids around, tool making continued. I recognize that this is argument by exclusion. I recognize too that chimpanzees can use tools, such as stones to crack open nuts. But there is a great conceptual leap from using stones as simple hammers with which to break things to using stones to strike a flake off another stone. The brains of australopithecines were not significantly bigger than those of modern great apes, including chimpanzees, taking body size into account. The brains of earliest *Homo* were significantly bigger, and that extra brain power must mean something.

The more parsimonious position is to suggest that stone tool making was the province only of *Homo*, from the earliest times in the record through to the end. Proving it one way or the other may never be possible. Those involved in the quest for human origins have to accept that some questions, no matter how germane, may never be answered.

One of the questions arises from the evidence that while *Homo* was opening up a new hominid niche, other hominids began to slip into evolutionary oblivion: they became extinct. Of the three or more species of hominid that existed between 2 million and 2.5 million years ago, only two made it through to a million years ago: *Homo erectus* and the robust australopithecine. And of these two, only one persisted beyond a million years ago: *Homo*. What can we say about this, beyond recording that it happened? Not much, and nothing for certain. But our intense curiosity about our ancestors forces us to wonder.

From our perspective as a living, successful species, we tend to regard extinction as a mark of failure. How often does one

169

hear of a doomed project being described as a dinosaur? Frequently. The dinosaurs died out sixty-five million years ago, so they must have been failures, mustn't they? Hardly, as they had already been around for some 160 million years. It just so happens that a concatenation of circumstances sixty-five million years ago—including, perhaps, Earth's impact with a comet—brought every dinosaur species to extinction. Bad luck, in other words. Throughout the history of life, effluorescence and extirpation of species groups has been a repeated pattern. As a result, 99 percent of all species that have ever lived are now extinct. The australopithecines are among them.

We can, I think, go a little further than dry statistics, salutary though they might be. First of all, the period from ten million years ago to the present has been a biological disaster for the world's apes. At one time there were twenty or thirty species of so-called Miocene apes in Africa, and just a handful of Old World monkey species. The passage of ten million years has seen the position reverse, with many ape species going extinct and monkeys diversifying wonderfully. There were several causes, including the loss of ape habitat through climate change. In addition, the monkeys apparently outcompeted the apes on their own ground by becoming more efficient fruit eaters. What of the hominids?

The initial evolution of hominids was part of the Miocene apes' response to cooling climes: some went extinct, but one of them became bipedal, to thrive where an ape could not live. Further cooling, around 2.6 million years ago, promoted further evolution to two extremes: the robust, small-brained, large-cheek-teeth species, and the small-cheek-teeth, large-brained species. If these two adaptations had not occurred when they did, it is quite possible that the hominid family would have come to an end completely between one and two million years ago. Extinction, after all, was the fate of hominid species like *Australopithecus africanus* that did not adapt to the extremes.

It is perhaps easy to see why *Homo* persisted, having become distinct from apes in its adaptation. But why should the robust australopithecine not have continued its initial success? After all, its specialization seemed well adapted to drier climes. If we can read the fossil record correctly, *Homo erectus* and the robust australopithecines occupied similar territory: near water sources, with a mosaic of wooded and open country, more tolerant than other hominids had been of arid environments. But *Homo*'s diet seems to have been so much broader that direct competition is out of the question. The meat component enjoyed by *Homo erectus* distinguished it clearly from the robust hominid.

I see no reason that bands of *Homo* would not have killed and eaten robust australopithecines when they could, just as they killed and ate antelopes and other prey animals. In fact, it is possible that the *Zinjanthropus* cranium my mother found at Olduvai Gorge in 1959 was an item on the menu of *Homo habilis*. If the cranium had been that of an antelope in the middle of an obvious living floor, we would have no hesitation in concluding it was a food item. Why not that of another large primate? But there is no evidence—such as cut marks in the bone—to suggest that the *Zinjanthropus* head had been worked on with stone tools. The suggestion must remain in the realm of logic and speculation. If there were evidence of defleshing, I suspect that we would draw the wrong conclusion. I've no doubt it would be labeled "cannibalism"—wrongly, I'm sure.

According to my dictionary, a cannibal is one who eats its own species. Therefore, by definition, a *Homo erectus* who ate a robust australopithecine would not be a cannibal. Even more to the point, to a *Homo erectus*, a robust australopithecine would, I suspect, look like just another large animal. It is our anthropomorphic perspective that imputes human thoughts and feelings to anything that looks human, however superficially. I believe *Homo erectus* had a well-developed sense of self and a considerable language ability. But it would be very used to having robust

171

australopithecines around, behaving in a very different manner. I doubt that, when it killed a robust australopithecine, it would have feelings any different from when it killed an eland or a baboon.

We know that *erectus* was a highly successful species, able to extend its territory outside Africa by a million years ago. Such an expansion implies population growth, and this may well have included a push into robust australopithecine habitat. Caught between this and the simultaneous expansion of populations of baboons, the robust australopithecines may have succumbed to competition of a more basic sort: access to food sources. A million years ago, that double-edged competition became too stiff, and the robust australopithecines became extinct, breaking forever a living link with our ancestors.

We started this discussion with teeth and proceeded through facial anatomy, brain expansion, a major shift in our ancestors' social structure, an equally significant shift in diet, and on to the evolutionary oblivion a million years ago of the last remaining australopithecine species. In the enlarged brain, the newly emerging tool-making ability, and the beginnings of a hunting-and-gathering subsistence, we recognize hints of ourselves, our humanity. That, I believe, is significant.

Equally important is the way in which the interconnectedness of things is so cogently demonstrated here. I'm often asked why this or that happened in our history, and I know that a straightforward, resonant answer is expected. But evolution is rarely simple cause and effect. There are many variables in the uncertain mix: the climate, the local geography, a species' evolutionary heritage, the nature of other species in the community, and a measure of pure chance. So when I'm asked why tooth development slowed down in early *Homo*, my straightforward answer is because the childhood period became extended. But the real answer is: "This way lies humanity."

10

A Pendulum Swings
Too Far

When   one flies northeast out of Koobi Fora, on the east side of Lake Turkana, one quickly leaves the shore behind and heads for a sinuous line of hills, some twenty miles distant. These hills form part of the eastern margin to the Turkana basin, and they are broken here and there by river systems that drain the Ethiopian highlands to the north. Countless millions of tons of silt, sand, and gravel have been brought down by these rivers over the ages, inexorably building sediments that trapped slices of time within their layers. These sediments entomb clues to the questions we ask about our origins.

Fortunately for those of us who want to uncover those clues, the geological cycle of the Turkana basin in these parts changed direction about a million years ago, so now, where once there was deposition, erosion is at work. In a sense, this phase of erosion is running the geological clock backward, as seasonal streams cut through the ancient deposits, gradually exposing glimpses of time long past and long buried. The immediate effect, to the eye, is a harsh landscape of gray and brown erosion gullies in intricate and often bewildering patterns. It is a profile typical of badlands, and the semi-arid climate supports only the most sparse vegetation cover.

About twenty minutes after we take off from the Koobi Fora airstrip the topography below changes, and a north-to-south rise, the Karari Escarpment, is visible; it formed because the ancient river deposits here are harder than those to the west. Starting just below the escarpment is a seasonal river, Sechinaboro laga, whose twelve-mile course westward to the lake is lined with a luxuriant ribbon of bushes and tall *Acacia tortilis* trees. The snaking green, a typical riparian pattern, contrasts dramatically with the parched terrain.

Apart from the shade offered by the acacias, the Karari area is not a particularly inviting place. Drinking water is scarce, and the lake is too distant for the pleasures of swimming or the necessities of washing. But the Karari is archeologists' heaven, because of all the areas around Lake Turkana, it is one of the richest sources of ancient artifacts. For that the archeologists are more than happy to make do with a washing-water ration so meager that only teenage boys could possibly consider it anywhere near adequate.

"Wonderful, just wonderful. There's so much to be done here" was how my friend and colleague Glynn Isaac once described the Karari region. His tragic early death in 1985 put an end to his personal plans, but for Glynn's students and associates the Karari will always be special. It is the location of one of their most ambitious collective projects, inspired by Glynn. Located below the scarp and just a short walk from the shaded stream course, the site is known technically as FxJj50 for its map coordinates, but is usually called Site 50.

Every summer for three years in the late 1970s, Glynn and his team descended on Site 50 and carefully exposed a land surface on which, 1.5 million years ago, our ancestors walked. "Some sites you excavate simply because they are there," said

Glynn. "But Site 50 was different. Our aim was to try to answer very specific questions, some of which were about the science of archeology itself, though ultimately they were about the ways of life of early proto-humans."

A debate had been going on among archeologists in the 1970s, a debate about interpretation of evidence, about how much you can know of the past from the litter that was left behind. It may sound like an arcane academic exercise, but it developed into a frequently acerbic confrontation. Most of all, it concerned our perceptions of our ancestors. "Site 50 was going to help settle that, or so we hoped," said Glynn. I'm no archeologist, but here was an issue that touched the essence of what I seek from the past. I watched and waited with more than a little interest as Glynn and his colleagues rolled back the eons, slowly bringing to life a brief instant in our ancestors' existence, 1.5 million years ago.

From the top of the ridge in which Site 50 is located is a twelve-mile view westward to the lake, a waterless and erosion-ravaged terrain. "You have to remember that the modern land-scape bears very little resemblance to the landscape 1.5 million years ago," Glynn explained to a group of visitors in 1980, one of the last occasions he went to the site. "Back then you would have been standing on a more or less featureless floodplain, halfway between the lake to the west and the hills at the eastern margin of the Turkana basin. What we've done in excavating Site 50 is to open a little window onto that ancient floodplain, onto the lives of a few *Homo erectus* individuals."

A small seasonal river had meandered by this spot 1.5 million years ago, its banks lined with bushes and trees, not unlike the Sechinaboro laga where Glynn and his colleagues made their camp each year of the excavation. And, again like Glynn's team, the *Homo erectus* group had chosen the bank as the most hospitable place to be. "The sand would have been soft to sit on, the trees provided shade, and there were lots of lava pebbles for

175

making tools," explained Glynn. "For some short period of time these proto-humans used this riverbank, and left behind them what at first sight looks like an unpromising jumble of bones and stones." About two thousand fragments of bone and fifteen hundred pieces of stone, to be exact. Shortly after the *Homo erectus* group last used the site, the water level in the river rose and spilled gently over the bank, carrying fine silt and sand with it. Evidence of what had happened during the site's occupation was quickly buried, until natural erosion and the archeologists' zeal exposed it again.

The excavated area measures roughly twenty yards north to south and ten yards east to west, with neat angles here and there, the way archeologists like to do things. As the excavation progressed through three years of patient work, the fragments of bone and stone were collected, their positions in the earth carefully recorded on a comprehensive site plan. When an excavation like this is complete, an ancient land surface is exposed just as its occupants had left it. By this time in the archeologists' operation, all the ancient litter has been packed in boxes, possibly stored in a museum, so one has to rely on the scatter of dots and crosses on the site plan to visualize what the site once looked like. But at the request of a BBC television team that was filming Site 50 in 1980, Glynn and his team replaced all the fragments of bone and pieces of stone on the ancient land surface, exactly where each had been found, so as to re-create a visual impression of this ancient riverside spot, just as its occupants had left it. "It was a remarkable experience for me," said Glynn. "To be able to see it complete like that. Remarkable."

"An unpromising jumble of stones and bones" is how Glynn liked to describe such a collection, but even the untrained eye could see here that the bones and stones weren't randomly distributed. The northwest corner was where most activity had taken place, where one or two individuals had apparently sat and made stone tools, flakes, and simple choppers. There were fragments of hippo bone in that area too, and parts of a zebralike animal, giraffe, bits of an antelope the size of an eland, and a piece of catfish skull. Stone-tool making and bone processing had apparently gone on in the southeast corner too, but to a lesser extent.

In the mind's eye, I could begin to envision what had actually happened at this shady riverside camp 1.5 million years ago.

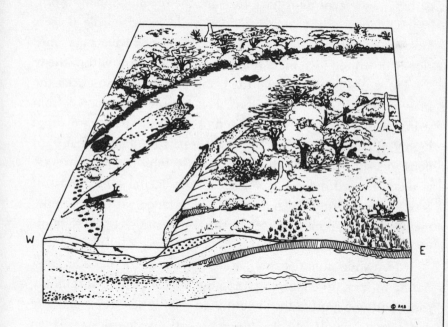

W   E

One and a half million years ago a small group of Homo erectus *people occupied a campsite on a riverbank. Reconstructed from geological evidence, the campsite looked something like this picture, and eventually was excavated by Glynn Isaac and his colleagues, to become Site 50. (Courtesy A. K. Behrensmeyer.)*

177

Here, on the opposite side of the lake from where the Turkana boy and his family were living at about the same time, a group of perhaps a dozen *Homo erectus* adults and children decided that the bank would make a good home base for a few days. The soft sand is comfortable to sleep on. The streambed, often dry, at this time is running with silt-laden water. The rainy season is coming, and distant storms can be seen and heard in the northern highlands. Meanwhile, the tall acacia trees that line the course of the stream provide welcome shelter, and the bushes carry sweet, succulent fruits that the children love to gather. The recent rains have brought the dry floodplain to brilliant vibrancy, with patches of tiny yellow and purple flowers looking like pools of color. The acacia bushes have burst into clouds of white blossom, belying the vicious thorns within them.

Three of the males leave the camp at first light to check the snares they set up the previous day. Simply constructed from strips of soft bark and sharpened sticks, the snares are often highly effective, trapping the leg of a passing animal that happens to snag them. The three males carry long sharpened sticks, as much for self-defense as for spearing a fleeing prey animal. It is not easy to bring down animals at a distance with such implements. Guile and persistence are the real weapons of these hunters.

Meanwhile, some of the mature females of the band are preparing to spend the morning foraging. Soft animal skins skillfully slung and tied around the shoulder do double service as papoose and carrier bag. After a few hours of work the females are laden with fruits, nuts, and succulent tubers, enough to sustain the entire band for the day. Like the men, the women also carry long sharp sticks for protection. They have shorter ones, too, comfortable to the grip, with which they dig for tubers. The foragers' skill is in knowing what fruits are ready for the picking and what wispy surface clues indicate the presence of nutritious tubers below.

Back at camp a couple of women and a man engage in idle

chatter, keeping an eye on the young child who did not accompany the hunting or foraging group. Yesterday, while he was stealthily tracking a young antelope, the man had slipped down a slope and gashed his leg on a jagged piece of lava. The hunt temporarily interrupted, his brothers tendered first aid. One of them searched for a clump of sansevieria, a succulent plant that grows around the lake. From some broken fronds, he twisted the juices, letting them drip on the open wound. The brother knew that unless he applied this natural medicine, the wound would become very red and the young man might die. Another stripped some thorns from a nearby acacia and began arranging them across the wound, piercing both sides of the gash. Thin strips of bark looped on alternate sides of the thorns drew the flesh together. First aid completed, the brothers continued on their way. Today, although the wound is sore, it looks clean, with little redness. The sansevieria worked, of course, he indicates to the women.

One of the women has been sitting at the edge of camp, deftly striking flakes from a lava cobble, the debris littered around her. She is now working strips of bark, softening them so that they will be supple and strong for making snares and tying skins. Another is using sharp flakes for whittling wood, making digging sticks. They wonder what the foragers—the hunters and gatherers—will bring back later that day.

The women, as always, can be relied on to bring back enough to keep hunger at bay; they represent the stable element in the economy. This day their haul is varied and plentiful, including some birds' eggs, probably flamingo. Soon, the hunting band can be heard returning, and from the noise they are making the camp knows that there will be meat to eat tonight. A large antelope was caught in one of the snares but had broken free, damaging its leg in the process. The hunting band spent much of the day tracking it and were able to kill the animal as it lay exhausted near the shore. While some of the men were dismembering the animal, another hunter noticed vultures near the lake

179

edge and went to investigate. He found a hippo carcass. Tomorrow the band will go back to see whether anything is worth scavenging. Today the antelope fulfills all their needs.

As always when meat comes into camp, there is great excitement: anticipation of the feast to come and reconstructions of the chase, sometimes more than a little embellished to add to the drama. One of the men looks for lava cobbles suitable for making sharp flakes, and soon has enough to butcher the antelope leg. Meanwhile, one of the children has speared a catfish in the nearby stream. As they share the fruits of the day's efforts, they agree that it is a good place to stay for a few days. Darkness falls, and lightning can be seen in the distant hills, too far for the thunder to reach the little camp.

There's a constant awareness of other, similar bands in the region, some of which contain relatives and potential mates. Young females in our group, when they reach maturity, will move to one of these other bands, and a network of relatedness and alliance will be built. Occasionally, the bands from distant parts that arrive in the region are sources of tension and some fear; physical aggression is possible when no alliances exist.

After some days at the camp, our band knows it is time to move on, not least because the storms are getting bigger. Thunder can now be frequently heard, and the rain in the hills is already swelling the stream. Soon it will flood the banks. The band can put off the move no longer, and they leave their camp, now a scatter of bones, broken stones, a catfish head, abandoned animal skins and tendons, ends of tubers too bitter to eat, strips of bark, and half-whittled sticks. The band will seek higher ground.

Not long after the band leaves the shaded camp, the silt-laden stream waters gently lap over the bank, slowly but inexorably covering the litter of a few days' existence. Some of that litter—the bones and stones—will remain, to become part of a rich archeological record. The perishable components—the skins, the tendons, the plant material—will decay, leaving a void in the archeological record that we try to fill in with deductions and informed speculation.

This scene seems to me a reasonable interpretation of what may have happened at Site 50, modeled on the ways of hunting-and-gathering people of today. What we are looking at here is something very human, something different from the way apes live. For humans, the food quest is often a cooperative affair, bringing resources to a home base, where it is shared. Not so for apes. As Glynn once remarked: "If you could interview a chimpanzee about the differences between humans and apes, including the way we walk, the way we communicate, and our subsistence, I think it might say, 'You humans are very odd; when you get food, instead of eating it promptly like any sensible ape, you haul it off and share it with others.'"

The sharing of food among members of hunter-gatherer bands is more than an economic transaction. It is the focus of complex social interaction, alliance formation, and ritual. If we are to be guided by ethnographic example, which is the most powerful source of inference in this case, then it is an eminently reasonable inference, in spite of modern, Western, feminist objections. Among both modern and historic foraging peoples a division of labor among males and females is important, with males doing most of the hunting while females are responsible for most of the gathering of plant foods.

Also important in the collective life of the earliest foraging people was an intensification of what Glynn called "social chess," a deep understanding and manipulation of other individuals' motivations and needs, social reciprocity. Social and intellectual sensitivities are raised to levels unreachable in the daily lives of apes. Communication is also more intense and sophisticated than ape vocalization. By *Homo erectus* times the range of sound production and the imposition of meaning and understanding would have been elevated to the point where they formed the rudiments of a spoken language.

There is no doubt that the adoption of the hunter-gatherer

way of life, with all the elements of humanness that go with it, was a key event in our evolution. Do we see signs of it at Site 50, 1.5 million years ago? Is our "reasonable interpretation" valid? Was *Homo erectus*—our Turkana boy—already to some degree human in his behavior? We've seen the indications from life history and developmental and anatomical evidence. What of the archeological record?

I cast the description of the activities at Site 50 in the context of the life ways of modern hunter-gatherers, more primitive in many respects, no doubt, but fundamentally the same. Until recently this was how most early archeological sites were interpreted, and this interpretation became the focus of debate among archeologists. "It seemed a very attractive interpretation," says Rick Potts, an associate of Glynn's, now at the Smithsonian Institution in Washington. "The home-base, food-sharing hypothesis integrates so many aspects of human behavior and social life that are important to anthropologists—reciprocity systems, exchange, kinship, subsistence, division of labor, and language. Seeing what appeared to be elements of the hunting-and-gathering way of life in the record, in the bones and stones, archeologists inferred that the rest followed. It was a very complete picture." Glynn noted that "to those engaged in the first round of research, this interpretation seemed common sense."

The idea that hunting and gathering—particularly hunting—was important in human evolution has a long history in anthropology, reaching back to Darwin. In his 1871 *Descent of Man* he wrote: "If it be an advantage to man to have his hands and arms free and to stand firmly on his feet, of which there can be no doubt from his pre-eminent success in the battle for life, then I can see no reason why it should not have been advantageous to the progenitors of man to have become more erect or bipedal. They would thus have been better able to have defended themselves with stones or clubs, or to have attacked their prey, or otherwise obtained their food." Here, the hunting of prey was considered part of the evolutionary package that shaped us physically, that drove the initial evolutionary wedge between us and the apes.

Once seeded, the idea of man the hunter took root, especially the image of man the *noble* hunter. It suited a need to see humans as somehow triumphant in evolutionary ascendancy over the apes, right from the beginning. In the 1950s Raymond Dart, the discoverer of the Taung child, portrayed our ancestors as hunters, but of a somewhat less noble character. In an essay titled "The Predatory Transition from Ape to Human," Dart characterized the human career as follows: "The blood-bespattered, slaughter-gutted archives of human history from the earliest Egyptian and Sumerian records to the most recent atrocities of the Second World War accord with early universal cannibalism, with animal and human sacrificial practices, or their substitutes in formalized religions, and with the worldwide scalping, head-hunting, body-mutilating, and necrophiliac practices of mankind in proclaiming this common bloodlust differentiator, this precarious habit, this mark of Cain that separates man dietetically from his anthropoid relatives and allies him instead with the deadliest of carnivores!"

Dart's ideas were popularized successfully by Robert Ardrey. In *The Hunting Hypothesis*, he encapsulated the kernel of the idea: "Man is man, and not a chimpanzee, because for millions upon millions of years we killed for a living." The phrase "the hunting hypothesis" was not Ardrey's but was borrowed from academia, where it described the prevailing theory of human origins. Not as imbued with vivid descriptions as Dart's writings, the anthropological literature of the 1960s and early 1970s nevertheless looked to hunting as the formative force in human evolution.

"To assert the biological unity of mankind is to affirm the importance of the hunting way of life," said Sherwood Washburn and C. S. Lancaster at a 1966 landmark conference titled "Man the Hunter." The title of this symposium was significant, they

183

said, because "in contrast to carnivores, human hunting, if done by males, is based on division of labor and is a social and technical adaptation quite different from that of other animals." The emphasis is on the social milieu of human hunting, a cooperative effort in which the gathering of plant foods is also important.

The force of the hunting hypothesis is easy to appreciate. It provided a plausible explanation of the crucial differences between humans and apes. And it offered anthropologists living analogues—modern hunter-gatherers—for the technical and intellectual capacities of our ancestors. By the mid-1970s, no one argued, as Darwin had, that hunting was part of the initial evolutionary divergence between humans and apes. For one thing, it was becoming clear that our ancestors were fully bipedal long before they began to manufacture stone tools. But the appearance of stone tools in the record, coincidental with evidence of expansion of the brain in the genus *Homo*, was taken as the beginnings of the hunter-gatherer way of life, the beginning of real humanity.

It was against this intellectual background that, in the mid-1970s, archeological assemblages were interpreted. When stones and bones were found together in ancient deposits, archeologists assumed they were looking at the remains of a hunter-gatherer camp. A primitive one, perhaps, but that was to be expected if it was two million years old. Man the Hunter, in embryo. A shift in the intellectual perspective was afoot, however, and the raw image of Man the Hunter was soon to be eclipsed.

First, objecting to the male dominance implicit in the hunting hypothesis, several anthropologists proposed an alternative, the gathering hypothesis. With social bias just as important as in the hunting hypothesis, but in the opposite direction, the gathering hypothesis argued that technology associated with plant-food procurement, together with the social bonds of females and

V*ervet monkeys, seen here in Amboseli National Park, Kenya, use a series of specific vocal warning signals, and engage in a great deal of soft vocalization in social situations. (Dorothy Cheney/Robert Seyfarth)*

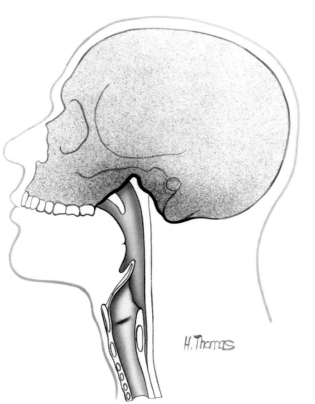

In humans the larynx is low in the throat, giving a large space in the pharynx in which a wide range of sound production is possible.

H. Thomas

The larynx in apes is high, and restricts the range of sound production possible.

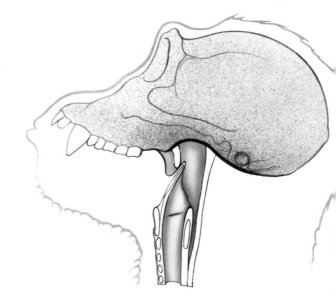

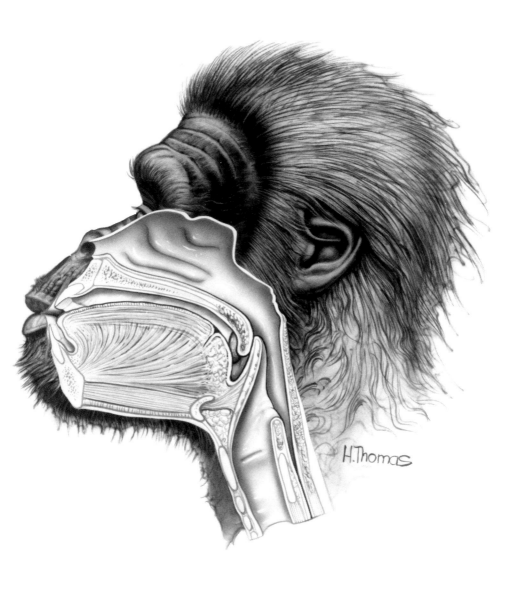

T his reconstruction of Australopithecus *shows the larynx in the ape position, implying a very limited range of sound production in these human ancestors. (H. Thomas)*

The excavation of the Turkana boy begins, with Alan Walker (left), Richard Leakey (middle), and Kamoya Kimeu (back). The thorn tree is still in place. (Virginia Morell)

Alan Walker displays the pieces of the boy's palate, just discovered, which indicate that there might be more fossil bone to be unearthed. (Virginia Morell)

T*en days into the excavation, work has pro-ceeded into the hillside. (Virginia Morell)*

A*lan Walker and Meave Leakey try to assemble the boy's cranium from fragments, a three-dimensional jigsaw puzzle. (Virginia Morell)*

Alan Walker excavates ribs and vertebrae, among some of the many anatomical elements of Homo erectus that were known first with the discovery of the Turkana boy. (Virginia Morell)

Samira Leakey, Meave Leakey, Alan Walker, and Richard Leakey pore over the Turkana boy's partly reconstructed cranium. A cast of an East African Homo erectus cranium is on the table. (Virginia Morell)

Mary Leakey examines the palate of the Turkana boy, in company with Richard Leakey and Alan Walker. (Virginia Morell)

Alan Walker, working at the museum in Nairobi, cradles the boy's cranium. (Michael McRae)

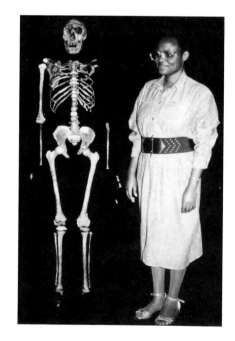

Emma Mbua, curator of the hominid fossils at the National Museum, Nairobi, stands next to the Turkana boy. Emma is 5'4" tall. (Alan Walker)

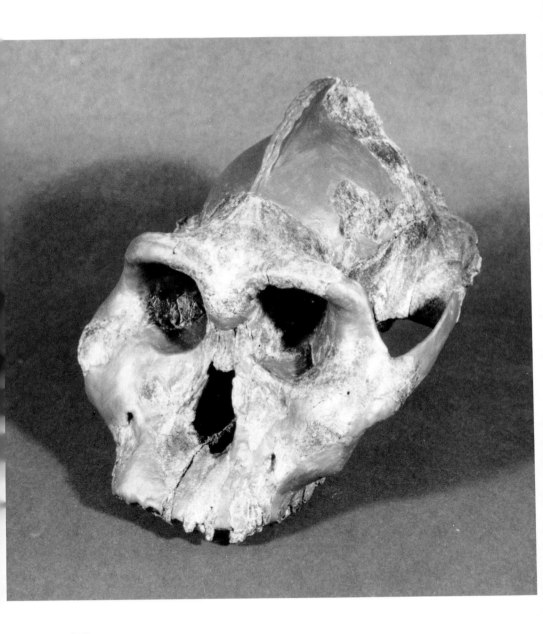

T he Black Skull, discovered in 1985, was a big surprise. It showed that very robust facial features had evolved early in australopithecine history, and may make the human family tree more complicated than has been imagined. (Alan Walker)

their offspring, was the underlying differentiator between humans and apes. Never widely popular, the idea nevertheless served to alert scholars to the way in which contemporary social values, male-oriented or female-oriented, may influence interpretations in anthropology.

Glynn was also moving away from the hegemony of the hunting hypothesis, choosing instead to emphasize the uniqueness of food sharing among humans. "The physical selection pressures that promoted an increase in the size of the brain, thereby surely enhancing the hominid capacity for communication, are a consequence of the shift from individual foraging to food sharing some two million years ago," he argued. He called his proposal the food-sharing hypothesis, and published it in 1978. It "includes division of labor among males and females, and a home base as a social focus for the exchange and consumption of meat and plant foods," he later explained. "I was quite explicit that, although meat was an important component of the diet, it might have been acquired by hunting or by scavenging. You would be hard pressed to say which, given the kind of evidence we have from most archeological sites."

Glynn's new position was very persuasive, and I wrote at the time, "The food-sharing hypothesis is a strong candidate for explaining what set early humans on the road to modern man." It combined the elements of a mixed economy one sees among modern hunter-gatherers without the macho image of Man the Hunter. Glynn's food-sharing hypothesis seemed consistent with what I could see in the fossil record, the archeological record, and what I considered biologically reasonable.

I was therefore alarmed—and bemused—when it became apparent that the abandoning of the extremes of the hunting hypothesis in favor of the more moderate food-sharing hypothesis was just the harbinger of change in intellectual perceptions in the archeological community. With a rapidly accelerating pace, ideas about early *Homo* ways of life successively shed elements of the hunting-and-gathering package. These proto-humans didn't hunt at all; they merely scavenged, it was said. They weren't even very efficient scavengers; they were marginal scroungers. Home bases were tossed out, and with them went the social milieu of food sharing, division of labor, and intensified social chess.

As I followed this process of dehumanizing our ancestors, I felt I was watching a pendulum swing from one extreme inexorably toward the other. Early *Homo*, according to some enthusiastic pushers of the pendulum, was recognizably human only in its upright locomotion, and in nothing else. Bizarre.

"We began to realize that our interpretations were strongly influenced by a set of unspoken assumptions," said Glynn, explaining how this process got under way. "It is surely true that by making and using stone tools, early hominids were departing from traditional apelike behavior. And if they were using tools to obtain significant quantities of meat, this would extend further that departure. But we realized that when we found tools and broken bones together we *assumed* there was a causal relationship, that hominids had been cutting meat from bones, breaking the bones for marrow, and so on."

So Glynn looked for a new archeological site at which to test these assumptions. "It's possible that broken animal bones could end up at a particular spot as a result of some kind of carnivore activity," he explained. "And it's possible that broken stones, artifacts, could end up at the same spot for reasons entirely unconnected with the animal bones. We had to test the idea that the bones and stones had been transported to the site by hominids. And we had to test the idea that the stones had been used to obtain meat from the bones."

Site 50 provided the opportunity for such a test.

11

The Human Milieu

While Glynn Isaac was beginning to re-examine his interpretation of early *Homo* behavior, Lewis Binford was sharpening his pencils in preparation for a vigorous foray into African archeology. Binford, an archeologist at Southern Methodist University in Dallas, is as noted for forcing archeologists to examine their methods of analysis and interpretation as he is for the acerbic tone of his critical attack. For instance, he suggested that researchers who interpreted early archeological assemblages in Africa as ancient living sites were "making up 'just-so' stories about our hominid past." It is an attention-capturing approach, but it irritated some scholars. It irritated me.

For a long time archeologists had more or less assumed that early *Homo* lived like modern hunter-gatherers, or at least a primitive version of that life. They assumed that when they found bones and stones associated in the archeological record, they were looking at the remains of hunter-gatherer campsites, which just happened to be 1.5 million or more years old. I'm convinced that Glynn was scientifically correct when he said we had to stand back from the process and try to sort out assumptions from valid interpretations. Site 50, which was to be part of that reexamination, to some degree got caught up in what is best described as the campaigning zeal that Binford created. Glynn, in his attempt to strip his interpretations of all unwarranted assumptions and speak only from direct inference, became too cautious.

Binford was skeptical of his colleagues' conclusions about ancient living sites because of his own experience with the record of Neanderthals, members of the human family who lived in Eurasia between about 135,000 and 35,000 years ago. He compared evidence of their way of life with that of modern hunter-gatherers. What he thought he saw obviously impressed him. "The more I learned about hunting and characteristic archeological signatures for typically human ways of life, the more I was convinced that ancient human beings—the Neanderthals—had been very different from us," he said. "If this was true, then the cozy picture of very early hominids painted by Leakey and Isaac for a much earlier time period appeared to be paradoxical."

If Neanderthals were as unskilled and unorganized in their subsistence strategies as Binford supposed, it is little wonder that he had difficulty in accepting that, almost two million years earlier, early *Homo* had already developed the rudiments of the hunting-and-gathering way of life. For Binford, the hunting-and-gathering mode is a recent development in human history. "Between 100,000 and 35,000 years ago the faint glimmerings of a hunting way of life appear," he concluded. "Our species had arrived—not as a result of gradual, progressive processes but explosively in a relatively short period of time." Binford posited that this late, explosive arrival of people like us was the result of the sudden invention of spoken language. My position is very different.

The life of hunter-gatherers has fascinated anthropologists

for more than a century, and in recent decades fine studies have been conducted on some of the rapidly dwindling number of foraging peoples. Not surprisingly, some aspects of the hunting-and-gathering life differ from one people to another, particularly when the environment is different. It would be unusual if, for instance, the Eskimos of the frozen north were to organize their lives identically with the San people of the Kalahari Desert. Nevertheless, some general patterns emerge, and these patterns imply a kind of inner coherence to the social and economic demands of hunting and gathering.

For instance, hunter-gatherers typically make a camp, or home base, in one location for several days, perhaps a week or two. They forage in the surrounding area, collecting the available plant foods and getting meat in any way they can. Then, when the resources begin to thin out, they move on to another location. There is a constant monitoring and exploitation of resources, with frequent moves to new areas. A campsite is rarely used more than once unless it is particularly rich in resources, such as the marine resources enjoyed by the Indians of the northwest coast.

Early *Homo erectus* bands must have used their landscape in many different ways too, so that one would not expect all archeological assemblages of bones and stones to be the litter of identically organized ancient home bases. But surely some of these assemblages represent home bases.

Not so, according to Binford. "The famous Olduvai sites are not living floors," he concluded in his book *Bones: Ancient Men and Modern Myths*. Not one of them. "The only clear picture obtained is that of hominids scavenging the kills and death sites of other predator-scavengers for abandoned anatomical parts of low food utility, primarily for purposes of extracting bone marrow," he said. Hence his characterization of proto-humans as "marginal scroungers." Archeological assemblages, according to Binford, were the result of proto-humans taking stones to abandoned carnivore kill sites, where they smashed a few marrow-bearing bones and then went on their way. Not a very human image.

And image is important, in science as in many endeavors, perhaps especially so in anthropology. Image affects the way in which evidence is interpreted: in this case, whether in the early archeological record we are seeing remnants of the activity of creatures that were humanlike or were far removed from humanity. To describe early *Homo*, including *Homo erectus*, as being "marginal scroungers" is to exclude them from other aspects of humanity. But if our ancestors were hunters of some skill, who lived complex social lives, then the consideration of other aspects of humanity—language, morality, and consciousness—becomes more acceptable. This is the philosophical crux of the so-called scavenger-versus-hunter debate.

One focus of research in this context for Glynn and his associates inevitably lingered on the bones at archeological assemblages. Assuming, as good evidence suggests, that hominids actually transported the bones to the sites, one asks whether the animals had been hunted or scavenged. When a hunter kills an animal, he may take back to camp whatever part of the carcass he desires. A scavenger is usually second in line to the carcass, and he can take only what the initial predator hasn't already eaten. The choice of body parts is therefore more limited for the scavenger. The patterns of bone assemblages at the home bases of hominid hunters and hominid scavengers would consequently be different. That's the theory.

"In practice it's very difficult to discriminate between a pattern that results from hunting and a pattern from scavenging," says Rick. It is sometimes impossible. "If a scavenger finds the carcass of an animal that has just died of natural causes, then all the body parts are available to the scavenger, and the bone pattern that results will look just like hunting. And if a scavenger manages to drive a predator off its kill very early, the pattern will again look like hunting. What are you to do?" It's a tough challenge, one that may elude all attempts to solve it. The Chicago anthropologist Richard Klein, one of the most experienced and thoughtful archeologists in the matter of bones assemblages, is not optimistic: "There are so many ways that bones can get to a site, and so many things that can happen to them, that the hunter-versus-scavenger question may never be resolved for hominids."

There are, however, tantalizing insights, such as that provided by one of the most unusual discoveries in this whole episode. For decades people had been talking about hominids as meat eaters and had analyzed the stones and bones collected from supposed home bases. Everyone assumed that the hominids had used the sharp stone implements to butcher carcasses. But no one had seen any direct evidence of the activity, such as the trace of a sharp edge on soft bone. Such cut marks can be seen in bony litter left behind by modern hunter-gatherers, whether they use blades of steel or of stone. But there was nothing in the archeo-logical record.

Then, in the summer of 1979, cut marks were discovered independently by three different researchers, all within a few months of one another. It was one of those remarkable coinci-dences in research, as if the right time had arrived for something new to be found. Rick Potts found cut marks. So did Pat Ship-man. And so too did Henry Bunn, a member of Glynn's team.

The marks appeared as short grooves, with a V-shaped cross-section, cut into the surface of fossil bone, from Olduvai and from Koobi Fora, preserving the activity of early Pleistocene butchers. Sometimes the cut marks were near the ends of bones, produced, presumably, when the proto-hominids were dismem-bering a carcass. Sometimes the cut marks were on parts of bones where only skin and tendons could be had. "For the first time, there was a firm link between stone tools and at least some

191

of the early fossil bones," said Pat, eliciting a collective sigh of relief from the archeological community. It was an important discovery, because, given the minimalist mental approach of some researchers, unless a direct causal link could be established between bones and stones, interpretations of archeological assemblages would be based on mere assumptions. I was delighted, but not surprised, to see the hard evidence of ancient butchery.

Even more evocative was the discovery that on some of the cut-marked bones were grooves left by carnivore teeth. Sometimes these ancient signatures overlapped, a cut mark crossing a gnaw mark, a gnaw mark crossing a cut mark. "When you see a cut mark on top of a gnaw mark, you can be sure you're looking at hominid scavenging," observed Pat. The carnivore clearly got to the bone before the hominid butcher. "Unfortunately, when you see gnaw mark over a cut mark, it's equivocal. The hominid may have killed the animal. But, then, the animal may have been dead when the hominid came along and butchered bits of it. Then a carnivore came along and had a gnaw at it. You just can't know for sure."

In the presence of this ambiguity, what can we say about carnivory in general that may be helpful? First, there are very few pure predators, like the cheetah, or pure scavengers, like the vulture. Most carnivores scavenge when they can and hunt when they must. I see no reason that our ancestors, when they became meat eaters, did not fit this pattern. I know from experience how easy it is to get meat as a hunter, and one doesn't have to be lethally armed to do it. Very effective are traps and snares, made from branches bound with bark strips, all of which would be invisible in the archeological record. As a boy I used to make crude traps of this sort. They are simple to construct and efficacious. I would be surprised if this kind of thing did not go back far into our history, probably beginning with the expansion of the brain in *Homo*.

. . .

What do we see when *Homo* comes along? We see a rapid expansion in brain capacity, and this must equate in crude ways with increased intelligence, which, among other things, would have enhanced technical skills. Even chimpanzees are successful at organized, cooperative hunts—catching monkeys collectively by cutting off routes of escape, for instance. But they are not known to make traps and snares for prey animals. I view it as probable that increased intelligence enabled early *Homo* to be even more effective cooperative hunters than chimps, and to make simple traps and snares.

But we see something else in early *Homo*, something that could well enhance our understanding of the life of these ancestors. It concerns the shape of the body, and derives from two different researchers with very different perspectives.

"We were sent a cast of the Lucy skeleton, and I was asked to assemble it for display," remembers Peter Schmid, a paleontologist at the Anthropological Institute in Zurich. For people interested in ape and human anatomy, the institute is a center of great renown. There, during the 1940s and 1950s, Adolf Schultz built up one of the world's best collections of ape skeletons. Schultz's work provided a foundation on which much of contemporary comparative anatomy is built, and his institute welcomes a constant stream of researchers who need to understand ape anatomy. "When I started to put the skeleton together, I expected it to look human," Schmid continues. "Everyone had talked about Lucy as being very modern, very human, so I was surprised by what I saw."

The chest was the problem. "I noticed that the ribs were more round in cross-section, more like what you see in apes," he explains. "Human ribs are flatter in cross-section. But the shape of the rib cage itself was the biggest surprise of all. The human rib cage is barrel-shaped, and I just couldn't get Lucy's ribs to fit this kind of shape. But I could get them to make a conical-shaped rib cage, like what you see in apes."

We know that Lucy had unusually long arms and relatively short legs, but the assumption was that, because she was bipedal, her body was like that of modern humans. After the experience with the rib cage, Peter decided he would look further into the anatomy of the entire upper body. He examined the whole trunk, the lumbar region, and the shoulders. He wanted to know how Lucy—*Australopithecus afarensis*—had moved about the landscape. Specifically, he wanted to know whether she was able to run bipedally, like humans.

The shoulders, the trunk, and the waist are important in human running: the shoulders for arm swinging and balance; the trunk for balance and breathing; and the waist for flexibility and swinging of the hips. "What you see in *Australopithecus* is not what you'd want in an efficient bipedal running animal," says Peter. "The shoulders were high, and, combined with the funnel-shaped chest, would have made arm swinging improbable in the human sense. It wouldn't have been able to lift its thorax for the kind of deep breathing that we do when we run. The abdomen was potbellied, and there was no waist, so that would have restricted the flexibility that's essential to human running." In other words, Lucy and other australopithecines were bipeds, but they weren't humans, at least in their ability to run.

While Peter Schmid was working with the Lucy skeleton in Zurich, Leslie Aiello was crunching numbers at University College, London. One set of numbers she was working with was the weights and heights of the San Francisco 49ers offensive line. "I needed data on some big, I mean really big, humans," she explains. They don't come much bigger than that. Most of her data were more conventional, however: the heights and weights of modern apes, and estimates of heights and weights of various hominid specimens, including Lucy.

194   She found a striking pattern. By comparison with humans,

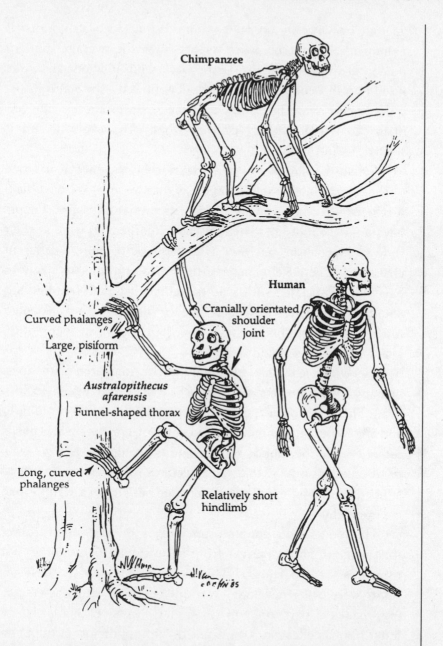

Chimpanzee

Human

Curved phalanges

Large, pisiform

Australopithecus afarensis

Funnel-shaped thorax

Cranially orientated
shoulder
joint

Long, curved
phalanges

Relatively short
hindlimb

The australopithecine species almost certainly were not adapted to a striding gait and running, as humans are. Here the adaptations to tree climbing can be seen in Australopithecus afarensis. (Courtesy of John Fleagle/Academic Press.)

apes are heavily built for their stature. For instance, a six-foot-tall chimpanzee might be twice as heavy as the average modern human of the same height. Leslie was asking how our ancestors fitted into this comparison. "No doubt about it," she states. "Australopithecines are like apes, and the *Homo* group are like humans. Something major occurred when *Homo* evolved, and it wasn't just in the brain."

Perhaps because we've been dazzled by the spectacular size of the human brain—mental egotists that we are—we have paid less heed to other aspects of our ancestors' physique. People have argued about how efficient australopithecines were at their style of bipedalism, but here is an analysis of quite a different character. "The development of the human physique is likely to have been associated with a fundamental change in hominid adaptation," concludes Leslie. Whatever bipedalism was adapted for in the first hominids, some important aspect of locomotion changed with the origin of *Homo*.

By having less bulk for their height compared with apes, humans have a greater relative surface area over which to lose heat. The long lower limbs give us an increased stride length, and the lower positioning of the center of gravity (in the pelvis rather than in the thorax) reduces inertia, or drag, when we walk or run. Leslie suggests that these features would be important to a bipedal hominid engaged in increased activity in a warm, open environment.

This conclusion conjures up images of humanness rather than apeness, and it derives from basic studies of anatomy, not from unrestrained fantasy. That is important. Just as these insights were being developed, a third pertinent idea was emerging, thanks in part to a trip to an auto mechanic and a chance letter from a colleague. Dean Falk, an anthropologist at the State University of New York at Albany, has studied the inside of fossil hominid crania for almost two decades, and is exquisitely familiar with the information to be gleaned from them. One thing she learned was that blood vessels that drained the brains of hominids fall into two patterns. In the earliest known hominids and most of the australopithecines, the blood flows in a few major vessels at the back of the brain and then into the jugular veins. *Homo* is different. The blood flows through a much broader network of veins, a pattern that becomes more elaborate through evolutionary time. This difference eventually led Dean to propose her so-called radiator theory.

Dean's conversation with her auto mechanic, Walter Anwander, a whiz who completely rebuilt her 1970 car, sowed the seeds of the theory. "One day, while he was enumerating the wonders beneath the hood," recalls Dean, "Walter pointed to the radiator and said, 'The engine can only be as big as *that* can cool.'" Hominids are machines of sorts too, and cooling is important, particularly for the brain. Not long after Walter Anwander's instructive observation, Dean published some of her results on cranial vessels in hominids. Michel Cabanac, a French physiologist, saw Dean's scientific report and wrote to her, pointing out that heat dispersal would become especially important as brain size increased in human history. Perhaps the drainage patterns Dean had discovered were pertinent to this.

Indeed, thought Dean, and she soon proposed the idea that the broad network of drainage veins in early *Homo* brains would have permitted strenuous, heat-producing activity. Not so in the early australopithecines. How to test the theory is difficult, if not impossible, to imagine. Nevertheless, it is at least consistent with the inferences drawn from body structure in australopithecines and *Homo*, and consistency is sometimes all one can hope for in scientific theories, particularly those dealing with historical events.

I was delighted when I heard on the grapevine about these ideas, and about Leslie Aiello's and Peter Schmid's results. It seemed to me that here was an angle that would catch the archeologists by surprise. Early *Homo* looks to me like a creature adapted to broadening its diet by becoming partly carnivorous in a physically active manner. It could run like us when it needed to; and it had tremendous stamina, like us. These are the marks of human hunters. Australopithecines had neither characteristic. They were not hunters.

But what of the notion that, as carnivores, proto-humans operated from temporary home bases, that a humanlike social and intellectual milieu was already emerging? We've seen that the period of childhood in *Homo erectus* must have been extended because of the immaturity of the infant's brain at birth. Is Site 50 an example of such a home base, or merely a conglomeration of bones and stones of little anthropological interest, as Binford has suggested?

The three years of patient work at the site paid off, yielding interconnected data of a quality and comprehensiveness never before achieved at an early archeological site. In the end, Glynn and his associates were able to demonstrate that the meat-bearing bones at the site had been transported there, almost certainly by the proto-humans themselves. Some of the bones bore the telltale cut marks; others had been broken open with hammerstones. The stone tools had actually been made on the site by one or more of the *Homo erectus* band. We know this because Ellen Kroll was able partly to reassemble some of the pebbles from which flakes had been struck. And in a clever microscopic investigation, Larry Keeley, of the University of Illinois, and Nick Toth, of Indiana University, were able to show that some of the stone flakes bore signs of use on meat, tough grass, and wood.

These various insights transform our image—my image, at least—of Site 50 from a place where desultory butchering took place to the site of many activities; in other words, a temporary hunter-gatherer camp.

Glynn was more cautious. Site 50 did demonstrate what he suspected: hominids of this era transported stones and bones to a chosen spot and used stone tools to get meat. But he stopped calling such places "home bases" and instead referred to "central place foraging." He said at a conference to commemorate the hundredth anniversary of Darwin's death: "My guess now is that, in various ways, the behavior system was less human than I originally envisaged." Glynn, I believe, was being too cautious, too much influenced by that pendulum. He said, "It is my strong suspicion that if we had these hominids alive today, we would put them in zoos, not in academies."

I am not suggesting that the Turkana boy was as human as we are today. But I challenge the notion that humanness arose very rapidly and very late in our evolution. I suspect that this extreme position has been adopted because of a desire to have ideas accepted in an unusual intellectual climate, an unconscious but powerful process. Many people believe that humans are so different from the rest of the animal world, they cannot accept the idea that we are a product of evolution, just like other species. Perhaps some anthropologists react to this unscientific position by emphasizing the special human qualities when they offer scientific explanations of our origins.

Much more reasonable, it seems to me, and much more consistent with the evidence, is the notion that qualities as complex as consciousness, morality, and ethics developed over a long period of time in our history. I believe that the Turkana boy lived in a rich social milieu, elements of which we would recognize as being human. And I speculate that when the Turkana boy died, his kin would have felt and shared emotions of grief much more like those which humans experience today than those which chimpanzees experience.

PART FOUR

■

In

Search

of

Modern

Humans

■

12

The Mystery
of Modern Humans

The fate of the Neanderthals is one of the most long-standing problems in paleoanthropology and one of the most current. Neanderthals first evolved some 135,000 years ago, eventually occupied a great swathe of Eurasia from Western Europe to the Near East, and finally disappeared thirty-four thousand years ago. Ever since their extraordinarily robust bones were discovered in the Neander Valley near Düsseldorf, in 1856, debate has ranged—and raged—over the place of Neanderthals in human history. Were they an extinct dead end, a branch of the human tree that died out? Or were they in some way ancestral to modern people in Europe?

Neanderthal anatomy was the focus of that lengthy debate. These ancient people were stockily built, with thick, powerfully muscled limbs. They were big-brained, slightly bigger than the average modern human, as a matter of fact. But the head anatomy was extraordinary. The cranium was long and low, with a "bun" at the back and protruding brow ridges at the front. And

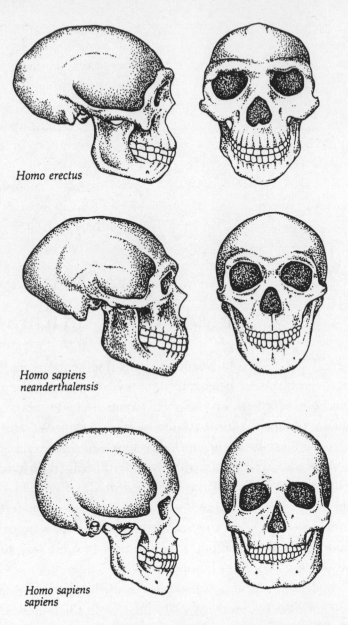

Homo erectus

Homo sapiens
neanderthalensis

Homo sapiens
sapiens

Although the Neanderthal brain was slightly bigger than the average brain of modern humans, the shape of the cranium was quite different, being relatively long and low. One of the most striking features of Neanderthals was the projecting mid-face, which does not appear in either Homo erectus or Homo sapiens.

204

the face was unique in human history. Imagine a modern human face made of rubber. Now take hold of the nose and pull it out several inches. The result is an oddly protruding central portion of the face, not just the nose but everything around it. That, roughly speaking, is a Neanderthal face. The question is this: Could such a striking anatomical form be part of our immediate evolutionary history and yet not be apparent in the physiognomy of modern European populations?

This question, once strictly in the purview of anatomy, is now being asked—albeit indirectly—in molecular genetics laboratories, particularly in the United States. And in January 1988, the results of the research were featured as the cover story of *Newsweek*, which gives some idea of the immediacy and drama of the new perspective. The molecular genetics approach considered the Neanderthal problem part of a larger inquiry, that of the origin of anatomically modern humans in general. Hence the somewhat provocative and very misleading title the editors of *Newsweek* chose for their story: "The Search for Adam and Eve."

According to the molecular genetics data promulgated in the magazine article, modern humans first evolved about 150,000 years ago, somewhere in sub-Saharan Africa. And what of Neanderthals? They were to be pushed to one side, an extinct human species that did not contribute to modern populations—no culture, no genes, nothing.

The effect of these assertions by molecular geneticists was to polarize the arguments over modern human origins—and, as a central part of it, the fate of Neanderthals—as never before. Perhaps it is because the data have to do with genes, not bones, that the rhetoric became so acerbic, both in public and in private. Certainly the fire was fanned by the fact that the chief proponent of the molecular data, the Berkeley biochemist Allan Wilson, chose not to hide his disdain for some of the anthropological opinions. As we saw, Wilson had cause for confidence in the molecular approach: he and his colleague Vincent Sarich were right once before with genetic data, when they claimed that humans and apes diverged about five million years ago, not the fifteen million years that we anthropologists favored for so long. The conclusions that Wilson and his colleagues make concerning the origin of modern humans are as radical now as the date for the ape-human split was back in 1967. Will they be proved right again?

Before we work our way into the folds of this complex story, we need to step back and view the big picture. We are concerned here with the origin of modern people, the final emergence of humanity.

Our canvas is large and varied in texture. In terms of evolutionary time, it covers the period in our history from some time before two million years ago until the end of the last Ice Age, about ten thousand years ago. In terms of anatomy, it traces the shift from an athletically built, muscular, bipedal, relatively large-brained creature (early *Homo* and *Homo erectus*) to a lithe, bipedal, large-brained creature (anatomically modern humans, *Homo sapiens sapiens*). In terms of technology, it tracks a change from a stable, modest tool assemblage of a dozen or so implements to a kaleidoscope of ingenuity, with hundreds of finely crafted designs coming and going with unprecedented speed. Ultimately, and most poignantly, it concerns the evolution of the human mind, and with it the sense of esthetics, the sense of morality, the sense of invention, and the sense of wonder about our place in the universe of things.

Some scholars have argued that the origin of modern humans was as significant as the emergence of the genus *Homo* itself. We are, they suggest, the product of a sudden cognitive efflorescence, which generated a modern level of spoken language and conscious awareness. If so, it could be viewed as the third of three great steps in human history: the origin of bipedalism; the origin of the big brain; and the origin of introspective consciousness. That progression has a cogent appeal.

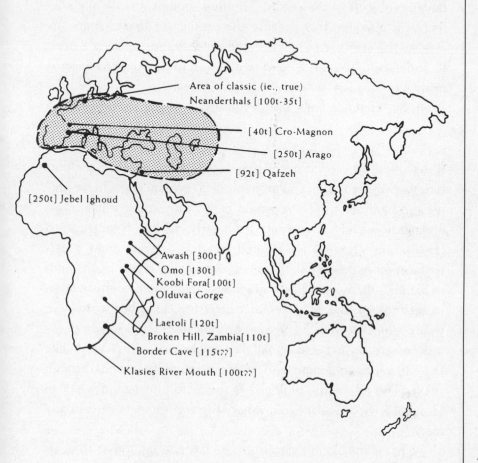

Area of classic (ie., true)
Neanderthals [100t-35t]

[40t] Cro-Magnon

[250t] Arago

[92t] Qafzeh

[250t] Jebel Ighoud

Awash [300t]
Omo [130t]
Koobi Fora[100t]
Olduvai Gorge

Laetoli [120t]
Broken Hill, Zambia[110t]
Border Cave [115t??]

Klasies River Mouth [100t??]

The shaded area shows the geographical extent occupied by classic Neanderthals, between 100,000 and 35,000 years ago. Other sites show early modern human remains, such as Klasies River Mouth in South Africa and Qafzeh in the Middle East; and transitional (archaic) forms, such as Arago in France, Jebel Ighoud in Morocco, and the Awash in Ethiopia. (The figures are dates in thousands of years.)

207

How are we to learn whether it is correct? Scientifically, we are restricted to hard evidence—the fossils, artifacts, and other tangible objects in the record—limited though it is. While correct, it is also true that because the product of that change—us —is such a creature of wonder and possessed of so deep a need to understand, there is a great temptation to go beyond the hard evidence. This is acceptable only if it is clearly recognized where scientific inference ends and desire-driven speculation begins.

If evolution proceeded in the way often imagined, then the picture gained from fossil and archeological evidence would be simple and harmonious. There would be a steady and gradual "modernization" of human anatomy from early *Homo* to *Homo erectus* to *Homo sapiens sapiens*—a less rugged skeleton, a larger brain, a flatter face, more delicate skull bones, and smaller cheek teeth. And, in parallel, there would be a gradually heightened sophistication in technology and in styles of expression. In other words, we would expect to see change marking measured and steady progress toward an expected goal. But evolution doesn't work like that; it varies in tempo and mode through time and through space. The job of the biologist is to try to understand what, in terms of any particular evolutionary history, the patterns actually mean.

What is the overall pattern in the last two million or so years of human history, and particularly in the last 1.6 million years, which witnessed the evolution and disappearance of *Homo erectus* and the eventual evolution of *Homo sapiens?* Regrettably, the fossil record of the period between 1.6 million years ago and the present is much less complete than anthropologists would like and, I suspect, nonanthropologists believe. The anthropological map of this time may appear to be densely populated with famous names like Java man, Peking man, Arago man, Heidelberg man, Solo man, Broken Hill man, Steinheim man, Bodo man, and many more. But in reality, the record of any particular geographical region is very patchy, so that tracking evolutionary change is often difficult. There are, for instance, no unequivocally recognized specimens of *Homo erectus* in Europe—anywhere. This incompleteness of the record throughout the Old World has, I believe, contributed to the current disagreement over the origin of modern humans.

We can say, however, that wherever *Homo* specimens are found from between 1.6 million and about half a million years ago, they can be labeled *Homo erectus:* tall, strong individuals, with relatively large brains, a fairly long cranium, thick skull bones, forward-jutting face, and prominent brow ridges. (There is no hint of the protruding midface so characteristic of Neanderthals.) Moreover, after about thirty-five thousand years ago, all that is ever found is *Homo sapiens sapiens*, modern humans. We therefore have to account for the period between half a million and thirty-five thousand years ago, no easy task.

There are many fossils from this elusive period, (including, of course, the Neanderthals), but the pattern is by no means simple, and the interpretations are as divergent as they can be. The easiest—and least satisfactory—thing to say about these fossils is that they appear to be neither one thing nor the other, neither *Homo erectus* nor *Homo sapiens sapiens*, but to combine elements of both. Neanderthals fall into this category, their big brains a claim to modernity, their robustly built skeleton and cranium the legacy of a more primitive past.

Much the same can be said about the Petralona skull, a large, very robust cranium that was discovered more than two decades ago in a stalactite-filled cave thirty miles southeast of Thessalonika in Greece. Petralona man is older than any Neanderthal, possibly dating to about 300,000 years ago. It has a big brain, about 1250 cubic centimeters, which is a hundred less than that of the average living human; its face is thrust forward less than in *Homo erectus* but more than in modern humans; its brow ridges are less prominent than in *Homo erectus* but more than in modern humans; the cranial bone is thick. A good mix of old and new, an apparent mosaic of features.

Likewise Arago man, a face and partial cranium from a cave in the foothills of the French Pyrenees. My friends Henri and Marie-Antoinette de Lumley have for some years organized extensive excavations in this spectacular cave, just a few miles from the small village of Tautavel. More primitive-looking than Neanderthal, Arago man nevertheless has that mosaic look about it, ancient and modern, ciphers of evolutionary ebbs and flows in our history. Henri and Marie-Antoinette speculate that the face may have been sliced off a skull at death and used as a mask in some kind of ritual. There is no way of knowing whether this was so, of course, but, entrancing though the notion is, proper scientific caution demands that I doubt it.

A dozen or so specimens display the same kind of mosaic look we see in Arago man and Petralona man, with regional features in Africa, Asia, and Europe. A few of their names are familiar, some less so. There is no question that something was going on in this period, ripples of evolutionary activity throughout the Old World. Because of the elements of modernity in these specimens—mainly the increase in the size of the brain and the less protruding face—they have long been known as "archaic sapiens." The idea was to recognize that they were on the threshold of "sapienshood" but not yet there.

Ian Tattersall dismisses the phrase archaic sapiens as "weaseling terminology," because, he says, "it is a way of avoiding the main issue." The issue he means is the need to decide what the mixture of ancient and modern features implies for our evolutionary history. "It's crazy to lump everything here into one taxon, especially if there is no real name for it," Ian says. "The term archaic sapiens is just a ragbag and has nothing to do with biological reality. Any mammalian paleontologist seeing morphological differences on the order separating modern humans from their precursors, and the latter from each other, would have no difficulty in recognizing a number of separate species."

Three species, maybe even four, he says. I am a champion of "the more hominid species the better," as I said earlier. I have no difficulty in envisaging several hominid species coexisting, as Old World monkeys do today. And I expect that in the period of two to three million years ago, there were more coexisting hominids than we now recognize. But I must admit to balking at the idea of three or four human species coexisting a few hundred thousand years ago, at the threshold of *Homo sapiens sapiens*. We are dealing here with humans, people very close to us, people who, if we met them, would be virtual mirrors of ourselves.

"The urge to include forms as diverse as Petralona, Steinheim, Neanderthal, and you and me in a single species *Homo sapiens* must be sociological in origin," counters Ian. "The only rational explanation for the taxonomic corraling of these widely differing fossils is the setting of an unconscious 'cerebral Rubicon,' perhaps somewhere at around 1200 ml. One can only applaud the generous liberal sentiment that leads to the inclusion in *Homo sapiens* of all hominids whose brain size falls comfortably within the modern range," he adds, with a touch of sarcasm. Am I a "generous liberal" to believe that these large-brained fossils do indeed qualify for the sapiens appellation? I think not. The large brain is the product and the engine of evolution in this late stage of our history, and to ignore its significance in this population of fossils is to be ungenerously conservative, to follow Ian's metaphor.

The overriding issue here, however, is how this perplexity of anatomical form fits into the larger pattern, the evolution of modern humans. There are two interpretations. At one extreme is the notion of a strong evolutionary continuity through time and space, an inexorable evolutionary force leading from *Homo erectus* to archaic sapiens to *Homo sapiens sapiens*. Wherever populations of *Homo erectus* became established in the Old World, this

211

model argues, *Homo sapiens sapiens* would eventually emerge via an archaic sapiens intermediate (including Neanderthals in Europe), and they would interact with one another through contact and gene flow.

This is known as the multiregional model, and fifteen years ago I described it with the following analogy: "Take a handful of pebbles and fling them into a pool of water. Each pebble generates outward-spreading ripples that sooner or later meet the oncoming ripples set in motion by other pebbles. The pool represents the Old World with its basic sapiens population; the place where each pebble lands is a point of transition to *Homo sapiens sapiens;* and the outward-spreading ripples are the migrations of truly modern humans." It is a graphic illustration of the idea, and I've seen several versions of it recently as the topic has become more hotly debated. However, I am no longer as sure as I once was that it is correct.

One reason the model appealed to me was that it encompassed the idea of inevitability in human history, an evolutionary momentum that, once established, led irrevocably to humanity as we know it. This may sound like a belief in the predestination of humankind, a conviction that humans were meant to be. I know that many people feel it to be true, but this is not what I imply at all. It seems to me reasonable to consider whether the development of an elaborate material culture and an extensive social and intellectual communion might in our immediate forebears have changed the rules of evolution in subtle ways.

Specifically, I suspect that a positive feedback may have built up, in which this kind of social and intellectual milieu would promote its further development. In other words, culture became an active component of natural selection, enhancing it still more. My father used to say that, through culture, humans effectively domesticated themselves. As we know, domestication—of plants and animals—leads to rapid evolutionary change. By analogy, the emergence of fully modern humans may have been accelerated through the effects of culture. If this is correct, then evolution toward fully modern sapiens would have occurred wherever the evolutionary momentum had established itself—in other words, multiregional evolution.

But new fossils and new dates for old fossils have begun to convince many of my colleagues that the multiregional model is incorrect. I don't believe that the fossil record has produced the final word yet, but it is suggestive enough to force even the most tenacious multiregionalist to look at the alternative, the second interpretation. What does it offer?

The idea here is that, instead of evolving wherever *Homo erectus* became established, *Homo sapiens sapiens* originated as a single evolutionary event—a speciation—in a geographically discrete population. Fully modern humans then spread out from this geographical region, replacing existing premodern populations (including Neanderthals) throughout the Old World. In a classic paper on the topic a decade and a half ago, William Howells of Harvard University called this model the Noah's Ark hypothesis. Since then it has also been denoted the Garden of Eden hypothesis, and, with the more recent molecular genetics evidence, the mitochondrial Eve hypothesis.

The two models—the multiregional model and the Noah's Ark hypothesis—could hardly be more different, both in the assumed evolutionary mechanisms underlying them and in the predictions for the fossil record.

The multiregional model, for instance, describes the evolutionary transition of one recognizable species into another (*Homo erectus* into *Homo sapiens sapiens*, via intermediate stages, the archaic sapiens) throughout Africa, Asia, and Europe. How likely is this, in terms, say, of current ideas on population genetics? "Even under ecologically identical conditions, which rarely exist in nature, geographically isolated populations will diverge from each other and eventually become reproductively isolated," observes Shahan Rouhani, a geneticist at University College, London.

Given the different ecological conditions that prevailed in Africa, Asia, and Europe at the time—and prevail now—reproductive isolation was very likely among *Homo erectus* and archaic sapiens populations in the Old World, he implies. "It seems to me that the multiregional model of modern human origins is therefore theoretically implausible."

The University of Michigan anthropologist Milford Wolpoff, the current standard bearer for the modern version of the multiregional model, disagrees. "There was considerable gene flow between the populations," he argues. "Gene flow is the latticework that connects the populations." Thus united, geographically distant populations can evolve in concert, if not precisely simultaneously, then along a common trajectory. That trajectory is drawn, Milford suggests, by the increasing elaboration of culture among our ancestral populations. In a sense, proto-human culture is not only a product of our ancestors' behavior; it is also part of the selection pressure that drives further evolution. This is a notion with which I have a lot of sympathy.

I have to admit, however, that the notion of culture-driven, multiregional evolution is quite outside what most geneticists understand about population biology. "Very large populations have a genetic inertia," explains Luigi Luca Cavalli-Sforza, a geneticist at Stanford University. "It would take a very long time for mutations to move through such a population. I don't see how the multiregional model could work." Did culture provide a commonality among premodern populations that unified them in such a way as to promote relatively rapid genetic change over a wide geographic area? I can't say, and at present I don't believe anyone can be dogmatic about it either way.

The alternative model—the Noah's Ark hypothesis—is different in terms of evolutionary mechanism and, for the most part, is much more consonant with most currently accepted population genetics arguments. Geographically separated populations of a species are seen as developing genetic diversity among themselves through time. Under these circumstances, a genetic shift—through mutation—may occur in one of these populations and convey an adaptive advantage over other populations. In that case, a new species may become established locally, one that will be reproductively separate from the original species. As far as biologists can determine, there might then be two closely related species where once there was one, and the two may continue to live side by side or come to occupy different territories. The Noah's Ark hypothesis is unusual in that it posits complete extinction—over a wide geographical area—of the original species, leaving the new species in splendid isolation.

Just as the underlying evolutionary mechanisms of the two models differ, so do the predictions of what the fossil record should look like. If the multiregional model is correct, then in each region of the world there should be a series of fossils showing increasingly modern features but with distinct and deeply rooted local differences. In other words, some anatomical characteristics of modern Asians would be detectable in archaic sapiens individuals of the region, and even in *Homo erectus* individuals of the region. The same for African characteristics and European characteristics.

According to the Noah's Ark hypothesis, by contrast, continuity of regional characteristics from *Homo erectus* times to the present are not to be expected. Instead, anatomically modern humans would be found first in one region of the world—the site of the speciation event. These modern forms would be found later in other parts of the world as they migrated away from the site of origin, replacing existing populations as they went. Looked at globally, there would be a single point of origin, from which a wave of modern humans would flow in all directions, sweeping all pre-existing human populations into evolutionary oblivion.

. . .

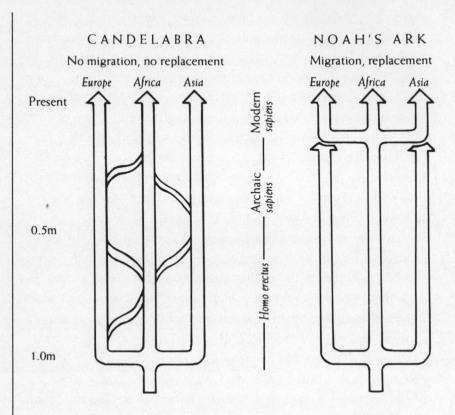

CANDELABRA
No migration, no replacement

NOAH'S ARK
Migration, replacement

Europe   Africa   Asia

Europe   Africa   Asia

Present

0.5m

1.0m

Modern *sapiens*

Archaic *sapiens*

Homo erectus

The two models have different implications for modern racial characteristics too. In the multiregional model, for instance, racial differences associated with different geographic groups would be the product of a long evolutionary history. They would be deeply rooted. In the Noah's Ark model, racial differences would be recent manifestations, the products of recent genetic differentiation. The roots would be shallow.

My friend Stephen Jay Gould is convinced that the Noah's Ark model is correct, and exults at its implications. "All modern humans form an entity united by physical bonds of descent from a recent African root," he wrote a couple of years ago. If there is one thing to be learned from a study of human origins, it is the unity of humankind. "We are one species, one people," I wrote a

216

decade and a half ago, in the book *Origins*. At that time, this was to me one of the most profound messages a study of human origins brings to a modern view of ourselves and of our future, for that matter. Even from a multiregionalist perspective, which I took firmly at that time, the message was unequivocal. If the Noah's Ark model is correct, and all modern people are descendants of a recent African population, then, as Steve says, our physical bonds of descent are yet closer. On an emotional level, I am, not surprisingly, strongly drawn to the Noah's Ark model. Its implications resonate with my convictions—and my hopes—for humankind. What does the hard evidence say?

One might expect an unambiguous answer to emerge from a study of the hard evidence, but this is far from the case. My anthropological colleagues are about equally divided in their assessment of the fossil record, half supporting a form of multiregional model, half a form of the Noah's Ark hypothesis.

Part of the problem is the incomplete fossil record. In most regions of the world there are large gaps in time. These gaps make it difficult, if not impossible, to establish secure evolutionary links between populations that lived a million years ago and those living today. In addition, the regional significance of individual elements of anatomy—the flare of the cheekbone, the shape of the forehead, the form of incisor teeth—is by no means agreed on. Some scholars argue that they can identify true regional characteristics. Others strongly dispute this.

Nothing would please me more than if the fossil record pointed firmly to one conclusion. Clearly, my colleagues do not think it does. And, try as I may, I am in no position to resolve it either: I simply do not see an indisputable signal. Under these circumstances, the proper scientific effort—in addition to looking for more fossils—is to turn to independent evidence.

Enter Mitochondrial Eve.

13

Mitochondrial Eve
and Human Violence

"*T h e M o t h e r* of Us All—A Scientist's Theory." This headline in the March 24, 1986, issue of the *San Francisco Chronicle* was the first most people heard of the idea that all of us can trace part of our genetic inheritance to a single female who lived in Africa some 150,000 years ago. Not surprisingly, the term "mother of us all" begged the biblical appellation African Eve, and the name has stuck, in scientific and nonscientific writing alike.

Eve was born out of molecular genetics data, particularly from Allan Wilson's lab at Berkeley. Here, in principle at any rate, was an independent line of evidence on the question of modern human origins, one that might prevail where fossil evidence so far has failed. We shall witness the birth of Mitochondrial Eve and see how she was received by the people whose business is fossils. And we shall see how the lines of evidence measure up against that other tangible signal in the prehistoric record: archeology, the indicator of our ancestors' behavior.

First, Mitochondrial Eve. Most genetics research focuses on chromosomes in the nuclei of cells, where most of the genetic information is packaged. However, a small number of genes are to be found in structures known as mitochondria, whose job in the cell is to produce energy. Two interesting properties of mitochondrial DNA make it particularly useful for tracking evolutionary history of recent populations. First, the DNA accumulates mutations rapidly, and therefore acts as a fast-ticking molecular clock. Second, because mitochondria are inherited maternally—from mother to offspring—they offer geneticists a relatively uncomplicated way of reconstructing evolutionary events in populations. By looking at patterns of genetic variation of mitochondrial DNA among modern human populations, anthropologists, theoretically, should be able to determine when and where the first members of anatomically modern humans evolved. This, in effect, would be the family tree of *Homo sapiens sapiens.*

Several laboratories took up this challenge in the early 1980s, including that of Douglas Wallace and his colleagues at Emory University in Atlanta, and Allan Wilson and his colleagues at Berkeley. Both labs produced a modest flow of scientific papers on the subject for several years. But not until the beginning of 1986 did the news hit the headlines. Almost a year after the *San Francisco Chronicle*'s "Mother of Us All" story, Wilson and his co-workers Rebecca Cann and Mark Stoneking reported in *Nature* that their mitochondrial DNA evidence supported a view favored by some anthropologists: "The transformation of archaic to anatomically modern forms of *Homo sapiens* occurred first in Africa, about 100,000–140,000 years ago." All present-day humans are descendants of that African population, they added.

In other words, the Berkeley team had seen that the overall degree of genetic variation in the mitochondrial DNA of modern populations is modest, which implies the relatively recent origin of modern humans. But of all existing populations, Africans have the deepest genetic roots, suggesting that an African population was the source of all other populations. This pattern fits the Noah's Ark hypothesis.

Because of the unusual dynamics of mitochondrial inheritance, the mitochondrial DNA in each living human being can be traced back to a single female, who lived in Africa over 100,000 years ago, according to the Berkeley team. Hence the popular term Mitochondrial Eve. In fact, it is somewhat misleading, because the single female from whom we all derive our mitochondrial DNA was a member of a population of humans at that time, not a lone mother. This population may well have been substantial, perhaps ten thousand people. The idea of Adam and Eve, literal mother and father of us all, is a headline writer's whimsy.

When Wilson and his colleagues published these conclusions at the beginning of 1987, they had data on just 145 individuals, representing several Old World populations—African, Asian, Caucasian, Australian, and New Guinean. By now they have checked more than four thousand individuals, with no diminution of their message. A key part of it is that when modern humans spread into other continents, they replaced existing populations, with no genetic mixing, no inter breeding.

Archaic forms of Homo sapiens in Asia would have been long established (back to Homo erectus), so their mitochondrial DNA would have accumulated large numbers of mutations. At least some of these mitochondria would have been incorporated into the migrating populations of anatomically modern humans if genetic mixing through interbreeding had taken place. Ancient Asian mitochondria would therefore be present—and be detectable—in modern Asian populations. "There is no evidence for these types of mitochondrial DNA among Asians studied," noted Wilson and his colleagues. "Thus we propose that Homo erectus in Asia was replaced without much mixing with the invading Homo sapiens from Africa."

Given his belief in the Noah's Ark model based on fossil evidence, Christopher Stringer welcomed the mitochondrial DNA evidence as strong support for his position. Just a year after Wilson's *Nature* paper came out, Chris and his British Museum colleague Peter Andrews published an article in *Science*, presenting both the fossil and genetic evidence in concert. It seemed a powerful combination. "Paleoanthropologists who ignore the increasing wealth of genetic data on human population relationships will do so at their peril," they warned.

The Stringer-Andrews paper was an important event in the heated debate, and it impressed many observers. Certainly I was impressed. It encouraged me to believe that the genetic evidence from Wilson's lab should be very seriously weighed. A year later, in February 1989, I was due to deliver the keynote address at a meeting on molecular evolution at Lake Tahoe in California. In the 1970s, I had been more reluctant than most to accept Wilson and Sarich's genetic evidence in favor of a recent (five million years ago) origin of hominids, so I thought this would be a chance to redress the balance. In the course of my talk I mentioned the mitochondrial DNA evidence and indicated that "I was ready to be persuaded by it." Surrounded as I was by molecular biologists and geneticists, I imagined it would be a wise thing to do, and scientifically proper too.

I was therefore more than a little surprised when, in the bar after my talk, several participants, including the conference organizer, Stephen O'Brien, cornered me and said, "You don't have to swallow that Mitochondrial Eve line. We don't." Steve and his friends proceeded to tell me why they thought the Eve hypothesis incorrect. Wilson wasn't at the meeting to defend his position, so I came away from Lake Tahoe with the potential pitfalls of the mitochondrial story, as argued by O'Brien and his colleagues, uppermost in my mind. Wilson may have miscalculated the rate of the mitochondrial clock; older mitochondria may have been lost by chance, promoted perhaps by occasional crashes in local population size; natural selection may have favored some recently evolved mitochondrial variant, thus eliminating the older lineages. Any of these possibilities might erroneously leave the impression of a recently emerged population, O'Brien explained.

These kinds of criticisms have been heard more and more since the Lake Tahoe meeting, promulgated particularly by Milford Wolpoff. In February 1990, Milford and half a dozen like-minded colleagues organized a session at the annual gathering of the American Association for the Advancement of Science, in New Orleans, the goal of which was to "nail this Mitochondrial Eve nonsense." Speaker after speaker argued for evidence in support of regional continuity and against localized speciation; for alternative interpretations of the mitochondrial DNA data and against the African Eve notion. It was a powerful presentation, and garnered a lot of press, with headlines like "Scientists Attack 'Eve' Theory of Human Evolution" and "Man Does Not Owe Everything to Eve, Latest Findings Say." Chris Stringer, who was speaking at a different session of the meeting, described the anti-Eve seminar as "high-powered salesmanship." One of Milford's assault team, David Frayer of the University of Kansas, summarized the deep reaction to Wilson's work: "Fossils are the real evidence."

As a "fossil person" myself, I can sympathize with that sentiment. When you look at fossils, you can see the anatomy, you can feel the morphology, and, if you have the right eye for it, you can strive to identify evolutionary relationships. Fossils, after all, are the tangible remains of what actually happened in our history. I have sufficient faith in our ability to interpret fossil anatomy to believe that—with adequate evidence—we shall be able to reconstruct past history. But I've also learned of the potential power of genetic evidence. Any anthropologist who chooses to ignore it or characterizes it as not real evidence does so at his or her peril, as Chris Stringer and Peter Andrews warned. At the same time, genetic evidence is subject to the kinds of uncertainties experienced by all sciences. And as long as the geneticists themselves are not unanimous about the validity of the mitochondrial DNA data, as expressed pointedly in February 1992, it seems wise to be cautious in accepting one interpretation of those data.

In the end the issue will be settled not by who shouts loudest or publishes most, but by the quality of the evidence itself. And, because the researchers are tracking the same piece of history, the fossil evidence and genetic evidence will in the end have to be consistent with each other.

It so happens that the consistency may already have begun to form, although it is still too early to tell. While Wilson and his colleagues were working up their mitochondrial DNA data, some European researchers were applying the technique of thermoluminescence dating to twenty specimens of burned flint from a cave site in Israel. The results were to give an approximate age of skeletal remains of anatomically modern humans that had been found in the cave Jebel Qafzeh, near Nazareth, between 1935 and 1975. Those results proved to be quite a surprise and to make the Eve hypothesis more likely.

The Middle East is interesting geographically, located as it is at a crossroads between Africa and the rest of the Old World. Descendants of a founding African population of anatomically modern humans, if such a thing indeed existed, would have passed through this narrow land corridor as they slowly spread north. Traditionally, however, the Qafzeh individuals were not interpreted in this out-of-Africa context. For many years they were thought to be about forty thousand years old, fitting neatly into the idea of local evolution from older populations. Those older populations were Neanderthals, remains of which are found in several cave sites in the area, including Kebara. The Kebara Neanderthals are about sixty thousand years old, so they

223

provided a population that, millennia later, evolved the more modern features of *Homo sapiens sapiens* seen in the Qafzeh people. Or so it was thought.

The new date for Qafzeh, published in February 1988, made this chronology impossible. According to the European group, which included Bernard Vandermeersch of the University of Bordeaux, the anatomically modern humans of Qafzeh lived more than ninety thousand years ago. That is, these people were in the area at least thirty millennia earlier than the Kebara Neanderthals. In that case, the Neanderthals cannot have evolved into anatomically modern humans, either in the Middle East or, for that matter, in the rest of Europe. Neanderthals disappeared from the Middle East about forty-five thousand years ago, some ten thousand years earlier than they vanished from Western Europe. Replacement of the Neanderthals by incoming populations of anatomically modern humans seems the most reasonable interpretation of events here. Note, however, how these arguments rest on dates of fossils, not on anatomical analysis. Change the dates, and you may change the evolutionary progression.

If anatomically modern humans truly occupied the Middle East around ninety thousand or more years ago, where could they have come from? "There are several candidates for modern human remains earlier than Qafzeh in sub-Saharan Africa," says Chris Stringer. "There's Border Cave, at perhaps 130,000 years, and Klasies River Mouth, at almost 100,000 years, both in South Africa. These dates are a bit uncertain, particularly Border Cave, but my guess is that we could be looking at an origin of modern humans somewhere in East Africa, with migrations north and south from there." There are some early modern human fossils in East Africa—in Kenya, Tanzania, and Ethiopia—though precise dating is a problem with most of them. But they do provide candidates for Chris's putative founding population.

Suppose modern humans really did evolve in East Africa

more than 100,000 years ago. And suppose they had reached the Middle East by ninety thousand years ago, where they coexisted with Neanderthals for as long as forty thousand years before moving farther north. How are we to explain this unusual period, when modern humans and Neanderthals coexisted for forty millennia? (A much shorter period of coexistence occurred in Western Europe when modern humans reached that part of the continent.) It is extraordinary to imagine.

One possibility is that the coexistence was more illusion than fact. The climate during this period of Earth history—the late Pleistocene—was unstable, driven by fluctuating glaciations. At times the Middle East would have experienced a temperate climate; at others, a much more changeable one. There is evidence of vegetational shift from temperate forest to pine forest and back again, all within the space of two thousand years. Although this may seem long by the temporal standards we are used to, it is but a brief interval in geological time. How could rapidly fluctuating climates create the illusion of many millennia of coexistence between Neanderthals and modern humans?

Neanderthals were a quintessentially cold-adapted people, as we can see in their chunky anatomy and in their occupation of northerly reaches of Eurasia, even during the coldest phases of the late Pleistocene. The first modern humans, by contrast, were quintessentially warm-adapted people, as is evidenced in their lithe build. Perhaps during the coldest periods the early moderns migrated south, vacating the Middle East. The Neanderthals, also drifting south, may have found the Middle East a suitable environment. During warmer climes, both populations would have followed their preferred environments north, the Neanderthals leaving the Middle East, the moderns entering it. Such a game of climatic musical chairs, some scholars argue, could have kept the two populations separate. Their only contact would then have been when each found remains of the others' occupation of their common range. Only when modern humans learned to survive and thrive in colder climes could they have pushed north into Neanderthal territory, which they began to do sometime after fifty thousand years ago.

Coexistence as illusion or reality? It is a difficult question to settle. There is no unequivocal evidence of interbreeding, such as the existence of hybrid individuals; no indications of exchange through contact or trade; no suggestions of violence, such as obvious trauma or, say, a Neanderthal individual having been consumed at the site of a sapiens camp. Our best answer lies perhaps in the technology, the tools with which each population interacted with its environment according to separate adaptations and habits.

About a million years ago, populations of *Homo erectus* began to move out of Africa and into various regions of the Old World, taking the Acheulean technology with them; that is, the ability to make large cutting tools, such as hand axes and cleavers. Throughout most of Europe and western Asia, therefore, where tool assemblages are found from sites older than about 250,000 years ago, they are typically Acheulean. (Tool assemblages in eastern Asia lacked Acheulean tools and looked very much like the chopping technologies of pre-Acheulean times.) The first major technological innovation came about 200,000 years ago. Archeologists call it the Levallois technique; it represents a leap in cognitive capacity, because it requires the preparation of a large, flat-topped core. There is a sense of "seeing" a desired shape within a rock in this case, not simply knocking a few flakes from a core. From the prepared core, different-sized flakes are then struck, each of which may be further trimmed. The result is an assemblage of as many as twenty or thirty different tool types, with points and edges and curves never before seen.

When you look at an array of these new tool types, you sense a different kind of mind at work, a qualitatively different interaction with the world. Overall, the Levallois-produced tools impart the strong impression of a set of designs. I can imagine that the complexity of design clearly imposed on stone, as evidenced in the tool kits, was also manifested as a greater complexity of thought, of language, and of social interaction. These mute stones, products of the Levallois method, bespeak a real change in the path of human history.

The Levallois technique actually became part of Mousterian assemblages, which began to appear at the same time in North Africa, Europe, and the Middle East. Named after the rock-shelter site of Le Moustier in the Dordogne, southwest France, Mousterian assemblages include backed knives, side scrapers, hand axes, denticulates, and points—all of which pushes the total range of tool types to about sixty. For the most part, Mousterian assemblages are associated with Neanderthal sites and are probably part of the overall Neanderthal adaptation. Elsewhere in the Old World, particularly in sub-Saharan Africa and parts of Asia, technologies of similar underlying skill appeared, produced by populations contemporaneous with Neanderthals. The technologies are known by a welter of local names too confusing to go into here.

All of this can seem very complicated, more so as one dwells on the details. But the overall pattern is simple: a technological innovation occurred around 200,000 years ago; it became the central part of a new, more sophisticated tool assemblage. Stability then prevailed once again, with no more significant innovations for at least another 100,000 years. Again, we see here an unimaginable—to us—period of lack of innovation. When innovation did come, it ignited a fire that burned with a recognizably human flame.

Slowly, and at first in Africa, the new technology emerged, one based on narrow blades rather than broad flakes. The blades were punched off a prepared core by the end of an antler or some such pointed device. The blades were then further shaped, to produce the finest of implements. The new technology was the beginning of a rush of stylistic innovation whose periods of change are measured in a few millennia rather than hundreds of thousands of years.

The archeology of this fascinating period is somewhat uncertain in the early stages, mostly because of a paucity of good, well-dated sites in Africa. Although blade tools are to be found among Mousterian assemblages, they are rare. In this new stage, which, through archeological convention, is called the Later Stone Age in Africa and the Upper Paleolithic in Europe, blades actually define the technology. Blades are the very essence of it. The signal appears first in Africa, a little less than 100,000 years ago, but its progress is difficult to trace in the few archeological sites currently known there. In Western Europe the task is easier, because of a long tradition of research and a wealth of suitable sites. The technology began there about forty thousand years ago and rapidly accelerated.

Although blades were the essence of the Upper Paleolithic, bone and antler were exploited as raw material for the first time, allowing extremely fine implements to be wrought, including eyed needles, used in making clothing. These industries were also associated with carving and engraving utilitarian and personal objects, and with engraving and painting cave walls.

We can say from this overall pattern that the origin of *Homo erectus* coincides closely with the Acheulean innovation; and that the escalating innovation of the Upper Paleolithic and Later Stone Age is associated with modern humans. But there is no clear archeological signal that unequivocally indicates the evolution of the earliest-known anatomically modern humans, no change in the archeological record that marches in lock step with the arrival of *Homo sapiens* as seen in the fossil record.

It is true that in Western Europe the appearance of modern tool technologies—the Upper Paleolithic—coincides with the appearance of modern humans, some forty thousand years ago. But almost certainly this represents the literal arrival of those people in that region, not a biological transformation on the spot. We do see the early development of the blade technology, which comes to define modern human technology, appearing in Africa close to 100,000 years ago. But it is a slow development, not a full efflorescence. To my eye this does not look like the kind of switch from presapiens to sapiens that we are seeking. And we know, of course, that in the Middle East we have modern people, such as Qafzeh, and a complete absence of modern tool technologies.

There appears, then, to have been an uncoupling of the evolution of the modern human form and the arrival of modern technology. This tells us, I believe, that it is a mistake to try inflexibly to link human anatomy and human behavior. The one certain thing we know about human behavior is that it is flexible, depending on physical circumstances and local mores. The earliest modern humans in one area may have organized their technology in one way; those in another area may have done something different. Different, but with the same underlying human cognition.

A second confounding factor is that the archeological record inevitably can provide only a minimum indication of what actually occurred. A crude stone-tool technology could be used, for instance, in the basic preparation of meat and plant food, simply for consumption against hunger. But the same set of tools could be used in food preparation that was part of an elaborate ritual. Yet the archeological record may not distinguish between the two events. Much of what is important in human social and ritual behavior is invisible in the archeological record.

The rich social traditions of Australian Aborigines, for instance, is expressed visually in materials such as feathers, wooden objects, colored sand drawings, and blood—none of which would be frozen in time as part of an archeological record. Nor would their songs, dances, myths, and ritual bodily incisions. We therefore must recognize that the archeological record is, at best, a minimal guide to the past, and especially to that part of the past in which we are most interested: the workings of the mind.

The question of the nature of the record begs another, which we need to address first: What kind of adaptive shift may have been associated with the origin of fully modern humans? Some of my colleagues argue that hunting strategies became much more sophisticated at this time. Lewis Binford suggests that premodern humans were incompetent hunters at best, lacking the planning skills, technology, and organization of real hunters. Opportunistic scavengers is how he characterizes them. This is far too extreme a position for my liking, and I am more convinced by Richard Klein's evidence.

Through work at the University of Chicago and in the field in South Africa, Klein shows the premodern people were modestly competent big-game hunters, but not of the more dangerous prey animals, such as Cape buffalo, pigs, elephants, and rhinos. The Klasies River Mouth Cave people he studied from the Middle Stone Age preferred to concentrate on eland, a much less hazardous prey animal. At the Later Stone Age site of Nelson Bay Cave, however, there is evidence that the anatomically modern humans who lived there were capable of hunting the more dangerous prey, including the Cape buffalo and bush pig. Klein describes the change through time here as "a significant behavioral shift."

Unfortunately, such shifts are mostly impossible to demonstrate conclusively in this fascinating part of our history. Another explanation of the evolution of fully modern human, one long popular, is a final, decisive expansion of linguistic capacity. Allan Wilson promoted this view, and even suggested that the crucial mutation was located in the genetic material of the mitochondria, not in the nucleus, as most people have assumed. The importance of the evolution of spoken language to the notion of humanity will be discussed more fully, but I make two observations here. First, it is inconceivable to me that the rapid increase in brain size we see in the evolution of the genus *Homo* would not in some way reflect a growing ability for spoken language. Second, it is equally inconceivable to think of a human species that has a linguistic capacity identical with ours, and yet is not completely modern, completely human. So, yes, an incremental and significant enhancement of spoken language may well have been part of the final evolution of modern humans, whether in the multiregional pattern or through local speciation.

If we view the evolution of fully modern humans as involving a behavioral and cognitive adaptive advantage, we are forced to face that putative long period of coexistence with Neanderthal populations in the Middle East. The subsistence skills of the Neanderthals would surely have compared favorably with those of the Qafzeh people; otherwise they would more rapidly have been swamped in a tidal wave of competition over resources. If the humanity gap was in fact quite narrow, then it makes even more poignant the notion of two such populations living as neighbors for so long a period.

The lack of evidence for genetic hybridization—interbreeding—may be taken to imply considerable differences between the populations. Perhaps a behavioral chasm effectively prevented mating; perhaps unions between the two populations were infertile. We have no direct evidence. We do know that, once coexistence came to an end in the Middle East, populations of people like us quickly spread throughout Europe and Asia. Coexistence between established populations and the newcomers in these continents was brief, perhaps a millennium or two. Again, there is no firm evidence of interbreeding between established Neanderthals and the newly arrived modern humans. But we know that they met: the Chatelperronian technology tells us that.

For years the Chatelperronian was a mystery. Discovered in

western France, it is a curious mixture of the typical Neanderthal flake technology and the modern human blade technology, including bone and ivory objects, and was therefore considered an intermediate technology employed by people in evolutionary transition between Neanderthals and modern humans. This interpretation was in the tradition of an evolutionary continuity between the two species. However, Bernard Vandermeersch and François Lévêque's 1979 discovery of the two Neanderthal individuals with the Chatelperronian assemblage at the St. Césaire rock shelter in western France effectively finished that idea. Neanderthals, it seems, had adopted some of the tool-making techniques of the newcomers to their land.

As there is no clear evidence of genetic mixing through interbreeding between Neanderthals and anatomically modern humans in this area of Western Europe, my guess is that contact—and technological exchange—would have been in the context of trade. Trade between technologically primitive tribes in the modern world is usually accompanied by the exchange of women, often in the context of building political alliances. In fact, this pattern of the dual trade of goods and brides was also common in historical times, among postagricultural communities. However, I do not find it difficult to imagine transactions of material trade in the absence of bride exchange. Neanderthals and Cro-Magnons were so different from each other physically that perhaps neither wished to be physically intimate with the other, even though the exchange of materials may have been acceptable. If, as I suspect, anatomically modern humans were superior linguistically to archaic populations, then communication between Neanderthals and Cro-Magnons would have been limited at best.

Perhaps communication was limited to some sort of ritualistic exchange of ivory pendants and fine artifacts. Perhaps this is how the Neanderthals became aware of a greater range of technology than theirs. Perhaps—almost certainly, in my view—this is one of those questions to which we shall never know the answer. In any case, the Chatelperronian technology spread throughout central and southwest France and northern Spain, and lasted for just a few thousand years. It was like a guttering flame, the last remnants of premodern human life before modern *Homo sapiens sapiens* became pre-eminent.

This brief coexistence in Western Europe raises the question of how it came to an end. Did the Neanderthals succumb to resource competition or to violence? If the Eve hypothesis is correct, the same question—competition or violence—would pertain to all the territory throughout the Old World that modern humans went on to occupy, in which they found established populations of archaic humans.

"Rambo killer Africans, sweeping through Europe and Asia" is how Milford Wolpoff characterizes—or, rather, caricatures—it. "You can't imagine one human population replacing another except through violence," he asserts. Given the lamentable history of the past few centuries—in the Americas and in Australia, for instance—violence perpetrated by newly arrived populations on existing populations at first seems a reasonable guess. The near genocide of American Indians and Australian Aborigines was in the tradition of colonial occupations with a long history of established warfare. Is it logical to infer similar genocide in the ancient past? Not necessarily.

The archeology of warfare fades fast in human history, rapidly disappearing beyond the Neolithic, ten thousand years ago, when agriculture and permanent settlements began to develop. The monumental architecture of the earliest civilizations seems often to be almost a celebration of warfare, of victories over the enemy. Even in earlier times, between five thousand and ten thousand years ago, indications of a preoccupation with military strife are to be found, often in paintings and engravings. But go back beyond that, beyond the beginning of the agricultural revolution, and the depictions of battles virtually vanish. I take this to be significant in the evolution of human affairs. I believe that warfare is rooted in the need for territorial possession once populations became agricultural and necessarily sedentary. Violence then became almost an obsession, once populations started to grow and to develop the ability to organize large military forces. I do not believe that violence is an innate characteristic of humankind, merely an unfortunate adaptation to certain circumstances.

The absence of indications of intergroup violence before the agricultural revolution does not of course *prove* that our hunter-gatherer ancestors earlier than ten thousand years ago were not as violent and as inclined to genocide as they have been in recent times. As always in science, the absence of evidence cannot be taken to be the evidence of absence. But I take it to be a very reasonable inference. Not reasonable, in my opinion, is Milford Wolpoff's assertion that because humans have been genocidal in recent times, they must have been so earlier. If it could be demonstrated that violence was the only possible mechanism for the replacement of one population by another, then we would be left with no explanation but Milford's. But that is not the case.

"For several years, I have been interested in modeling demographically the extinction of Neanderthals," Ezra Zubrow explained to a gathering of archeologists at the University of Cambridge. Zubrow, an anthropologist at the State University of New York, was attending a major conference on the origin of modern humans, held in the summer of 1987. Using computer models of population dynamics, he investigated the "interaction" of neighboring populations, given varying degrees of differences in competitive ability. His message was as clear as it was surprising. "I believe I can show that only a small demographic advantage is necessary for the modern forms to grow rapidly and for the archaic forms to become extinct." In the European context, he said, "the Neanderthals could have become extinct

234

in a single millennium." Which is precisely what we see in the record.

It seems counterintuitive to believe that a modest difference in subsistence skills—amounting to about a 2 percent margin in mortality per generation—could lead one population to success and the other to extinction. But often in biology our perceptions are rooted in present experience, and we have little grasp of what a long time dimension can do. In this case, a slim margin in mortality over a millennium translates into a big difference in ultimate survival.

Zubrow does not say, nor do I conclude, that modern humans did indeed outcompete Neanderthals. What his work shows is that interpopulation competition for resources is a plausible explanation of Neanderthal extinction over the kind of time scale with which we are dealing. The possibility must be taken seriously. Extinction through violence or through resource competition remain competing hypotheses until direct evidence unequivocally supports one or the other, or something else entirely. It is too easy to favor a particular hypothesis simply because it suits one's hopes for history or one's scientific turf.

If all of this appears to be a confusing and uncertain picture of modern human origins, it is precisely because anthropologists and archeologists themselves are not yet sure what actually happened. Much as we should like to know the answers to this major period in our history, we are certain only of the questions. But even among these questions, some are more easily formulated than others; and this probably means that some are going to be more easily answered than others.

For instance, we can reasonably hope that a combination of fossil and genetic evidence will one day settle the questions of when and where anatomically modern humans first evolved. Less certain will be the determination of the relationship between anatomically modern humans and modern human behavior, particularly in the context of stone-tool industries and artistic expression. Least certain of all is an understanding of the precise evolutionary change on which modern humans—the complete essence of humanity—is founded.

■

In

Search

of

the

Modern

Human Mind

■

14

The Loom
of Language

When we contemplate our origins, we quickly come to focus on language. Objective standards for our uniqueness as a species, such as our bipedality and our relatively enormous brain, are easy to measure. But in many ways it is language that makes us feel human. Ours is a world of words. Our thoughts, our world of imagination, our communication, our richly fashioned culture—all are woven on the loom of language. Language can conjure up images in our minds. Language can stir our emotions—sadness, happiness, love, hatred. Through language we can express individuality or demand collective loyalty. Quite simply, language is our medium.

Thomas Henry Huxley, Darwin's friend and champion, was greatly impressed with the importance of human language, and in 1863 wrote the following passage: "No one is more strongly convinced than I am of the vastness of the gulf between . . . man and the brutes . . . for he alone possesses the marvelous endowment of intelligible and rational speech [and] . . . stands raised upon it as on a mountain top, far above the level of his humble fellows, and transfigured from his grosser nature by reflecting, here and there, a ray from the infinite source of truth." Huxley is surely correct in identifying a language-created gulf that separates humans from the rest of nature. The capacity of *Homo sapiens* for rapid and detailed communication, and for richness of thought, is unmatched in today's world. The challenge for anthropologists, however, is to formulate the right questions about the origin of these skills.

Two major issues dominate the origin of language. First is the matter of continuity. Is spoken language merely an extension and enhancement of cognitive capacities to be found among our ape relatives? Or is spoken language a unique human characteristic, completely separate from any cognitive activities in apes? The second issue is function. Did language evolve as a tool of enhanced communication? Or did evolution select a less obvious ability, mediated by language?

Some modern linguists prefer to explain human language as an evolutionary innovation peculiar to *Homo sapiens*, having nothing to do with communication in higher primates. These linguists point to what is called the deep structure of language, particularly grammar, as evidence of human uniqueness in this form of communication. The way children learn language, and the overall similarity of structural rules in all languages, indicate a kind of language-acquisition device in the human brain. Noam Chomsky, the noted linguist at the Massachusetts Institute of Technology, has been instrumental, since the 1950s, in developing this now-dominant view of language.

I, like all parents I'm sure, was charmed and amused by the simple "errors" that my children made when they were learning to talk. Errors like incorrectly generalizing plurals, so that sheep were "sheeps" and mice were "mouses" or "mices." Similarly with past tenses of some verbs, like "taked" and "wented." Such errors are in fact instructive, because they demonstrate that in acquiring language children are not merely learning from what they hear adults say; they are generalizing rules, clicking in with their language-acquisition devices. Intimately related to this is the notion that because humans are one species, *Homo sapiens*, there will be an underlying structure common to all languages. The differences we see among the many languages of the world—indeed, the countless number of languages that must have existed through prehistory—are but variations on a basic structural theme.

If humans do possess a neurological apparatus that underlies language acquisition, grammatical structure, syntax, then the species may well be unique. This would be true if primate communication was not connected in any sense with what we humans call language and if primate cognitive capacities contained no hint of language competence. What is the evidence? It comes in several parts: first, from observations of primates in their natural setting, with an emphasis on their natural communication; and second, from investigations in primate research centers, the famous ape-language studies.

For many years now field observations of vervet monkeys have indicated that they produce three distinct alarm calls, a different one each for snakes, leopards, and certain eagles. When a vervet sees one of these predators and makes the appropriate call, the other members of the troop respond instantly and appropriately. At a snake alarm, they stand on their hind legs, look around in the grass, and then either mob the snake or scamper for the safety of trees. At an eagle alarm, they look upward and run for safety into the bushes. Eagles are able to take vervets both in trees and on the ground. So we see here a set of specific calls that elicit specific and behaviorally appropriate responses.

Now, when a vervet makes a *chutter* noise—the snake alarm call—it is not the exact equivalent of the word "snake" in a human language. We can use the word snake in many different contexts and many different abstractions. A vervet cannot. For this reason, a number of linguists have argued that primate calls should not be viewed as precursors of human language. This criticism might have some merit if the call-response system were absolutely inflexible, but even this is debatable. As Dorothy Cheney and Robert Seyfarth have shown in recent years, vervets' abilities are much broader than had been imagined.

These University of Pennsylvania researchers, having studied vervet monkeys in Amboseli National Park in Kenya for more than a decade, have shown that the monkeys are able to modify their use of alarm calls in subtle ways, depending on the precise circumstances. On one occasion they saw an eagle swooping to attack a monkey that was feeding on the ground. Several mature males saw the bird just as it was about to strike. Instead of giving the eagle alarm, which would have caused the intended victim to gaze up to the sky and then head for the bushes, they gave a leopard alarm call. The "wrong" call sent the individual running for the trees, just as it would have done if the warning was about a leopard. The animal survived the attack, but would not have done so had an eagle alarm been given and the appropriate response been followed.

Cheney and Seyfarth admit that this kind of event is rare, but suggest that the alarm system is more flexible than may be imagined. In addition, vervets use a series of grunts and other sounds in social interactions, exchanges that apparently convey a good deal of information about the current social circumstance. "Where once it was thought that vervets had a grunt, rhesus macaques a scream, and Japanese macaques a coo, we now know that the monkeys themselves perceive many variants of these signals, each with a different meaning," say Cheney and Seyfarth. "There is undoubtedly an upper limit on animal vocal repertoires when compared with the infinite number of messages that can be conveyed through human language. However, the size of vocal repertoires . . . is considerably larger than initially believed and the information conveyed by each call is less general than had been imagined." In other words, vervet "language" is not so far removed from rudimentary human language.

Most laboratory-based language studies in nonhuman primates have been carried out with apes, of course, particularly chimpanzees. Take Kanzi, for instance.

Kanzi is a male pygmy chimpanzee, born in 1980 at the Language Research Center of Georgia State University, in Atlanta. It just so happened that Matata, Kanzi's adopted mother, had been selected as a subject to learn a sign language, one of several studies on pygmy chimpanzee cognitive abilities at the center. Sue Savage-Rumbaugh, the researcher in charge, had developed a system of several hundred lexigrams, each of which had a specific meaning, like *run, milk, Kanzi*. By incorporating the lexigrams in conversation, pointing to them, and saying at the same time in English what they meant, she hoped to teach Matata to use them.

Ape-language studies have been under an academic cloud for a while, partly because some researchers were less rigorous in their interpretations than they should have been, and partly because the learning situations were usually artificial. The Atlanta work is now acknowledged to be among the best there is.

Even so, after many months of patience with Matata, Savage-Rumbaugh made little progress. Matata just wasn't learning much. One day, someone noticed that Kanzi seemed to understand some of the requests and instructions that were being made to Matata through the lexigrams and orally. "At first I didn't believe it could be true," says Savage-Rumbaugh. "But we started to test Kanzi actively, and sure enough, he had learned a lot of words, just picked them up as he played around while we worked with Matata." That Kanzi had learned words without formal instruction, mainly just watching and listening—as human infants do—is remarkable. "After that we started to work with Kanzi, and we gave up on his mother."

By now Kanzi has a large vocabulary and can respond to

such complex instructions as "Go to the bedroom, get the ball, and give it to Rose [a colleague of Savage-Rumbaugh]." Kanzi can do this, even when there is a ball in front of him at the time, which could be a source of confusion about which ball to get. "His comprehension is well developed," says Savage-Rumbaugh, "and to me that is an important aspect of what we think of as the substrate for language." Not only is Kanzi's comprehension impressive; his word production is well developed too, though not to the same extent. "We have demonstrated that a pygmy chimpanzee—a species virtually unstudied before from the point of view of language—has not only learned, but also invented, grammatical rules that may well be as complex as those used by human two-year-old children," says Patricia Marks Greenfield, a colleague of Savage-Rumbaugh, at the University of California, Los Angeles.

Savage-Rumbaugh believes that critics of earlier ape-language studies set unrealistic standards for assessing the significance of the work. "They seemed to say that, unless apes can do precisely what humans do in terms of language, then apes would fail the test," she explains. "That is simply not realistic. Language is a process of comprehension and word production. Grammar flows out of that. We know that Kanzi comprehends a great deal, and that tells us a lot." It tells us that in ape brains, which are the same size and organization as the brains of *Homo* ancestors, the cognitive foundations on which human language could be built were already present.

This does not prove that those same foundations were present in our ancestors, but to me it is highly suggestive. It tells me that the "vastness of the gulf between . . . man and the brutes" is not as great as many people believe. The sentiment of the "specialness" of *Homo sapiens* derives, I'm sure, from the sense of wonder that emanates from human consciousness and self-awareness. But we are in danger of being tricked by its cogency. The

evidence from primate language studies suggests that we are not as special as we would like to believe. For me the issue of continuity is clear: our language skills are firmly rooted in the cognitive abilities of ape brains.

We now move to the second major issue of language origins, that of function; specifically, whether spoken language evolved as a tool of enhanced communication, or for some other ability.

Human communication through language is unprecedented in the natural world, both in terms of rate and density of information transferred. The human vocal apparatus can produce about fifty different sounds, which, compared with about a dozen for the most vocal of animals, doesn't look impressive. But from those fifty sounds, or phonemes, the average individual can assemble a vocabulary of 100,000 words and an infinite number of sentences. A stronger argument could hardly be adduced in favor of communication having been the function that evolution honed through the emergence of spoken language. Or so it seems. Sometimes, however, the most obvious answers are not the correct ones. And in this case, the obvious answer has been strongly challenged.

Harry Jerison, of the University of California at Los Angeles, has made a special study of brain evolution throughout the animal kingdom, including humans. He concludes that expanding language capacity was responsible for the threefold increase in brain size during human evolution; and that greater language skills resulted from our need to build mental models in our heads, not primarily as a means of better communication. Jerison places his interpretation in the context of brain evolution throughout the animal kingdom.

In the history of life as a whole, we see an interesting pattern concerning relative brain size, superimposed on the pattern of evolution of major new groups, from amphibians and reptiles to mammals. At each step there is a dramatic jump in encephalization, an increase in the size of the brain relative to the size of the body. For instance, with the origin of mammals there was a four- to fivefold increase in relative brain size over their ancestors; a similar expansion occurred with the origin of modern mammals, fifty million years ago. In other words, each major evolutionary innovation was accompanied by a major increase in the amount of brain power. Presumably, that heightened brain power was in some way associated with the ability to function in the new survival niches.

If we are guided by the jump in encephalization we see in history, being an archaic mammal appears to have been more demanding than being a reptile or an amphibian. The same with modern mammals compared with archaic mammals. Although all modern mammals have relatively bigger brains than any reptile, not all mammals are similarly endowed. Primates, for instance, on average are twice as encephalized as the average mammal; monkeys and apes are twice as encephalized as the average primate; and relative brain size in humans is three times that of the average monkey or ape. Among primates, therefore, we see an increasing cognitive capacity among a class of animals—the mammals—that, by comparison with amphibians and reptiles, is already highly encephalized. What does it mean?

"Reality is a creation of the nervous system," explains Jerison. "The 'true' or 'real' world is specific to a species and is dependent on how the brain of the species works. This is as true for our own world—the world as we know it—as it is for the world of any species." What the brain produces, therefore, is a kind of mental model of the world, a system for handling the information that flows from sense organs and for generating the appropriate responses. The integration of the sensory data, one with another, is central to monitoring the world "out there" and to creating a model of it "in here." The "in here" becomes the real world as an individual animal experiences it.

If relative brain size is any guide to cognitive capacity—as it surely must be—then we can say not only that the real worlds of amphibians, reptiles, and mammals are different from one another, but also that there is a greater completeness, or at least complexity, of mental worlds through these three classes. By the same token, the average primate's real world is more complex than the average mammal's; the average monkey and ape's real world is more complex than the average primate's; and *Homo sapiens* occupies a world that is truly its own. But why so many different worlds?

The most obvious answer is that each major evolutionary innovation, each new level of real world, has enabled the species to be in some way better, more efficient perhaps, than the preceding one. After all, we tend to equate bigger brains with greater intelligence, and greater intelligence with some indefinable superiority. But this is very much an anthropocentric view. There is no evidence that the mammals of today's world are able to exploit their niches more effectively than, say, those great reptiles, the dinosaurs. If mammals were indeed superior in their exploitation of niches in the world, then a greater diversity of ways of doing it—as reflected in the diversity of genera—might be expected. However, the number of mammalian genera that exist today is about the same as at any point during dinosaur history. No sign of inherent superiority here. The spectrum of niches occupied in the two eras is similar too. Progress or improvement, in the sense that we normally use those words, is not an adequate explanation for stepwise enhancement of mental models.

The answer, suggests Jerison, has to do with the building of sensory channels and with history. In amphibians, for instance, vision is a primary sensory channel, whereas in reptiles olfaction also becomes important. In mammals, auditory faculties are acute, as are olfaction and vision. In primates, particularly the larger primates, olfaction diminishes in importance, with the emphasis on stereoscopic and color vision becoming stronger.

When several sensory channels become part of a species' cognitive repertoire—olfaction, vision, and audition, for example—the various inputs have to be integrated with one another. This inevitably demands more mental machinery than would be required for a single channel. It is also true that if evolution were able to start from scratch each time it produced a major innovation, then the end result could be relatively simple. But evolution does not start from scratch; it builds on what already exists. As a result some systems tend to become more complex through time. This apparent trend is the effect of history in long-term evolutionary change. Brains got bigger at major evolutionary innovations, partly through the development of new sensory channels or the modification of existing ones (and their deeper integration), and partly through the building on existing machinery, a kind of evolutionary leapfrog.

In many ways, therefore, a reptile in the small herbivore niche would perform much as a mammal in the same niche. The models of the worlds they inhabit, however, would be different. The ever-bigger brains, the ever-greater cognitive capacity, that we see through evolutionary time is what we equate with growing intelligence. "Intelligence," says Jerison, "is a measure of the quality of the particular real world created by the brain of a particular species."

In all primates, particularly large primates, the real world seems to have expanded over the average mammalian world. Very probably the intense sociality and complex alliance formation that occur in higher primate groups underlie some of the need for this broadened cognitive power. The ability to weigh present events with history and with the future is essential in this kind of social milieu. The production of the inner real world for primates—their mental model of the world—therefore depends to a much greater extent than in other animals on the processing of information. Working out the complex social rules of primate life, and understanding where in the picture one fits, demands a lot of brain power. Here, then, we see a major shift from simply collecting information through sensory channels to an emphasis on assessing that information. This represents the primate's niche: the quintessentially social animal.

What can we say of the mental machinery of humans, three times greater than that in the large primates? Once again, there seems to be an obvious explanation: technology. "Technology has long been regarded as the driving force behind human brain expansion," says Jerison. Indeed, one of the most influential concepts of human origins during this century was encapsulated in a little book that Kenneth Oakley wrote in 1949: *Man the Tool-Maker.* Oakley, a major figure at the British Museum for many years, and the uncoverer of the Piltdown hoax, argued that not only did man make tools, but, effectively, tools made man. In other words, as natural selection honed the manipulative skills required for tool making, a bigger brain evolved, making us more human. The image is of a positive evolutionary loop, in which greater manipulative skills required greater brain power, which in turn permitted a more developed technology, and so on. This seemed a reasonable argument, not least because in many ways we see ourselves as highly skilled technological creatures.

Jerison makes a very astute observation concerning the hypothesis of man the tool-maker. "It seems to me to be an inadequate explanation, not least because tool making can be accomplished with very little brain tissue," he says. "The production of simple, useful speech, on the other hand, requires a substantial amount of brain tissue." Language, that's what we need big brains for, implies Jerison. There is no doubt that through human history the neurological basis of technological skills improved through natural selection. But language shouts to be recognized as the skill that makes us different from our primate cousins.

If we look at that trajectory of brain-size increase through human history, we could describe it as the result of an evolution-

249

ary package comprising three components. First, an element of rising manipulative capabilities. Second, further development of the social skills that are already developed in higher primates. And third, the steadily enhanced ability to speak—or, more particularly, the ability to think—using a complex, propositional language. I place most emphasis on the third element.

With the evolution of *Homo,* and the beginnings of a hunting-and-gathering way of life, many things changed. Enhanced, specific communication—a spoken language—surely would have been a survival advantage, no question about that. But, argues Jerison, "the role of language in communication first evolved as a side effect in the construction of reality."

Just as the visual apparatus was crucial to amphibians in the creation of the mental world; just as the neurological machinery for an acute sense of smell contributed to the mental world of reptiles; just as the earliest mammals improved their mental models of the world with highly developed hearing; and just as primate minds created elaborate mental models by integrating and sifting sensory information; so too have humans added a special component to the mental machinery that creates our special reality. That component, argues Jerison, is language. "We can think of language as being merely an expression of another neural contribution to the construction of mental imagery, analogous to the contributions of the encephalized sensory systems and their association systems."

Equipped with language, or, more specifically, the facility of reflective thought, our minds create a mental model of the world that is uniquely human, capable of coping with complex practical and social challenges. That mental model was the product of an emerging hunter-gatherer way of life. It involved a balanced relation with the resources of the environment and a complex, tightly constructed social and economic contract within human groups. It was recognizably primate in origin, but unprecedented in its degree of development. Its primary product was human culture, a mix of things material and things mythological, things practical and things spiritual: a uniquely human mental model of the world, woven on the loom of language.

15

Evidence of Mind

A b o u t   f i f t e e n years ago Ralph Holloway, an anthropologist at Columbia University, came to the museum in Nairobi to look at 1470, the large-brained cranium found, in 1972, by Bernard Ngeneo, of my fossil-prospecting team. Dated to a little less than two million years old, 1470, a *Homo habilis*, is the oldest, most complete skull of *Homo* we have. Ralph was studying fossil hominid brains—he's a paleoneurologist—looking particularly at the overall organization as compared with apes'. He was also looking for Broca's area, a small lump that is found on the left side, toward the front, of modern human brains.

Broca's area is a landmark of language capacity in the human brain, although a somewhat uncertain one. At the time Ralph first visited the museum, I had long believed that language had an early development in human evolution. My belief, I have to admit, was based largely on intuition, not on hard data. Ralph's visit marked the beginning of roughly a decade and a half of various lines of research that support this belief. I know I am at odds with many of my colleagues, who prefer to view spoken language as a recent, sudden evolutionary innovation, perhaps as recent as fifty thousand years ago.

The evidence that points to an ancient establishment of spoken language comes from three sources. First, anatomical evidence on the organization of the human brain and of the vocal tract. Second, on the most tangible of all products of the human mind, stone tools. Third, on some of the more abstract products of the human mind, including abstraction through art and in ritual behavior.

The gross architecture of mammalian brains follows a general pattern. The brain is divided vertically from front to back, into the left and right hemispheres. Each hemisphere is divided into four sections, or lobes, each responsible for different sets of functions. The frontal lobe handles aspects of movement and emotions; the lobe at the back, the occipital lobe, is the location of, among other things, visual functions; the side, or temporal, lobe is important in memory storage; and the parietal lobe, at the side and top, plays an important role in integration of such senses as hearing, vision, smell, and touch.

Although many functions can be localized in the brain, one of the remarkable features of this organ is that some functions, often important ones, defy precise locating. One of these is consciousness. No one has been able to point to a region of the brain and say, This exclusively is the seat of consciousness. Even the location of language facilities cannot be established 100 percent. For example, an individual may lose relatively large sections of his or her brain, with no apparent loss of cognitive functions, including language. This should give us pause when we consider the large increase in brain size during our evolutionary history: the building of a big brain clearly is not simply the cumulative addition of discrete functional units.

Compare human brains with ape brains, and some gross differences in structure become evident, the most obvious of which is the relative size of the different lobes. For instance, in humans the temporal and parietal lobes are emphasized, pushing the junction with the occipital lobe farther back. In apes, the occipital lobe appears bigger and the frontal lobe less pronounced than in humans. Very loosely, then, we can speak of an overall humanlike organization of the major lobes of the brain and an apelike organization.

Neurologists have identified two language centers in the brain, Wernicke's area and Broca's area, named after the nineteenth-century researchers who first discovered them. As with much neurological research of this kind, information about location of particular brain functions is gathered from victims of brain pathology or trauma. For instance, Carl Wernicke discovered that patients with damage to the upper back part of the left temporal lobe often had problems with language comprehension, sometimes speaking fluently but with no sense. Similarly, Paul Broca found that damage to the lower back part of the left frontal lobe (just by the temple) often produced difficulties in speech production, even though comprehension was intact. Modern researchers are still trying to understand how language is organized in the brain, and it is evident that the system is anything but simple. Multiple areas and pathways are involved, and precise location is often elusive.

One effect of there being so much neurological apparatus devoted to language faculties is that the left hemisphere is significantly bigger than the right. Even in left-handed people, the speech centers are often, but not always, located in the left hemisphere. This left-hemispheric dominance, as it is known, is also associated with handedness. Apes have one hemisphere bigger than the other, too, but the effect is not as marked as in humans, and there is no predominance either right- or left-handed in ape populations. Neurologists assume that left-hemisphere dominance in humans is the result of the evolution of language. Preferred handedness has gone along with it.

Human

Parietal   Frontal

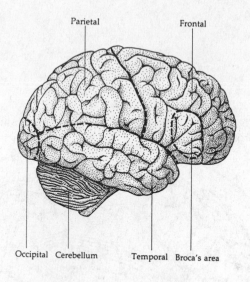

Occipital Cerebellum   Temporal Broca's area

Lunate sulcus Parietal   Frontal

Ape

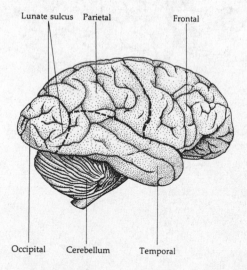

Occipital   Cerebellum   Temporal

*Although the overall structures of human and ape brains are the same, the proportions of the two hemispheres are different. In humans, the frontal lobe is large and the occipital lobe (at the back) is small. In ape brains, the occipital lobe is significantly larger, which pushes the lunate sulcus (the division between the occipital and other lobes) further forward. The position of the lunate sulcus is often taken as a "signature" of a human-like brain, and is therefore important assessing the status of fossil brain casts. (Courtesy of Ralph Holloway/*Scientific American, *1974, all rights reserved.)*

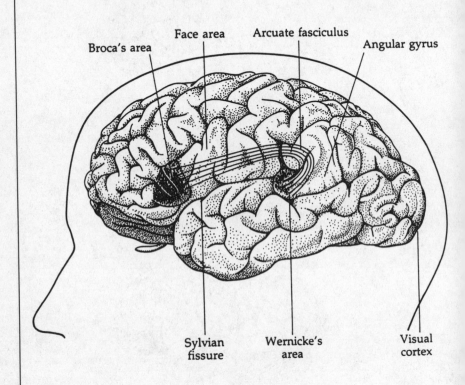

Broca's area Face area Arcuate fasciculus Angular gyrus

Sylvian fissure Wernicke's area Visual cortex

Two major language centers are recognized in the human brain, the Wernicke's area and Broca's area. Recent research shows, however, the language organization is much more complex than had been imagined. (Courtesy of Norman Geshwind/Scientific American, 1972, all rights reserved.)

In humans a distinct rise appears over Broca's area, a lump, a somewhat uncertain physical signal of the language abilities buried beneath it, but a signal nevertheless. This was why Ralph Holloway was looking for Broca's area in the skull 1470. If *Homo habilis* possessed a Broca's area of any size, it would have left its impression on the inner surface of the cranium.

"You do find Broca's area in 1470," concludes Ralph. "I'm not saying this proves that the individual had language, because in paleoneurology you can't actually prove anything. But I think that the origins of language extend far back into the paleontological past." In its earliest stages, human language presumably was an extension of vocal abilities, with a range of sounds and perhaps some structure to their expression. "The form of the language was undoubtedly primitive, but carried with it a limited set of sounds systematically used, and was based on a well-known aspect of primate sociality—the ability, if not the penchant, for making vocal noise." I strongly agree with Ralph's assessment.

If we wish to stretch the thread of inference yet thinner, we can look at the earliest stone tools for clues to brain organization. For many years Nick Toth, now at Indiana University, worked at sites at Koobi Fora as a key member of Glynn Isaac's team. Instead of excavating ancient stone tools, Nick often sought to understand past technologies by learning to make and use them himself. He is an experimental archeologist, with a quick eye for opportunities to test hypotheses. For instance, one summer while work was in full swing at a site near the Koobi Fora main camp, a big celebration was being organized by some local people, and a cow was slaughtered. Nick saw the occasion as a chance to test the efficacy of heat-hardened wooden spears. He quickly whittled some spears, annealed them in the embers of a campfire, and attacked the already dead beast. With enthusiastic encouragement from bystanders, Nick hurled the weapon at the animal's abdomen from close range. It bounced off. What ignominy, muttered Nick, and returned to his drawing board.

His more orthodox work is with stone tools. He saw that the way a stone knapper holds a stone pebble in one hand and strikes flakes from it with the other provides a clue to handedness. The clue is the shape of the flakes and the position of segments of the cortex, or outer surface, of the pebble on the flake pieces. Remarkably, Nick discovered that the pattern of flakes recovered from the earliest archeological sites in the Koobi Fora area is similar to the pattern he produces when he, a right-hander, makes tools. In other words, most of the early stone knappers—presumably *Homo habilis*—were right-handed. This surely implies that left-hemisphere dominance had already emerged by this time, a significant fact in the evolution of language.

Naturally, I was delighted with Ralph's discoveries of the Broca's area in the cast of 1470's brain, and with Nick's less direct but equally intriguing results on an ancient bias toward left-hemisphere dominance. A scientist always likes to hear evidence that supports his or her preconceptions. (An unfortunate corollary of this is a degree of deafness to evidence that goes the other way.) The presence of Broca's area cannot be taken as certain proof of language abilities, because in modern humans the linguistic machinery is buried beneath this neurological lump, not in it. At best, the presence of Broca's area is indirectly indicative of language abilities. Nevertheless, I was encouraged by Ralph's positive conclusion that 1470 had a degree of spoken language ability considerably in excess of ape vocal communication.

Fully developed spoken language like ours? No. I believe language abilities emerged gradually in the human career and were part of an evolutionary package built around the hunting-and-gathering way of life, something novel in the world of primates. Primatologists have developed a healthy respect for the degree to which monkeys and apes know their respective worlds. These animals know what to eat and, more important, where and when to find what's ready to eat. Even their daily meanderings across their ranges are seen as a good deal less than random. Sometimes a troop of baboons will split in two at the beginning of the day. The parties forage for food in different parts of the troop's range, and, at the end of the day, meet again to settle down to sleep, a troop once again united. Clearly, there is communication and "agreement" taking place here; the day's activities are planned in some as yet inexplicable way.

With a shift toward a mixed economy of hunting and gathering, in which daily divisions of the troop are routine in the separate food quests, the need for organization and agreement becomes yet more intensified. A sophisticated degree of communication is important, as it is for an overall increased sociality. If we compare a large locality containing many troops of baboons with a similar locality inhabited by many bands of modern hunter-gatherers, we note a key difference. Among the baboon troops there is an intermittent exchange of males as the youngsters near maturity. But there is never a grand coalition of all the troops, no temporary social aggregation. Among modern hunter-gatherers, however, such temporary aggregations of bands is almost a defining characteristic. The periods of aggregation are times of intense socializing, of renewing and evaluating alliances, of searching for marriage partners. This kind of aggregation is part of the tribal structure, bands united by a common language and a common culture.

Would we expect to find such a social organization among earliest *Homo*? I think not. We have to be cautious not to fall into the trap of imagining that once certain characteristics of humanness emerged in our ancestors, *all* characteristics were present. I suspect that in early *Homo* a rudimentary hunting-and-gathering system began to develop, accompanied by rudimentary language, but that the primate pattern of troops persisted for some time, perhaps until the evolution of archaic *Homo sapiens*.

The next question is obvious. If *Homo habilis* had a form of spoken language, what of earlier hominids? Here the evidence is less clear and is quite controversial. First, a distinct Broca's area of the sort seen in 1470 has not been found—yet—in an australopithecine. Is this proof of absence of language? Perhaps. But ever since his earliest studies on hominid brain casts, Ralph has argued that in all hominid species, australopithecines and *Homo*, a reorganization from the ape form has taken place, and the overall shape is distinctly human. "From the very beginning of the hominid lineage, a degree of humanlike brain organization had already been established," he says. If this is the case, rudiments of linguistic abilities may have been buried in the circuitry of the brain. But paleoneurology can deal only with surface features.

Ralph's suggestion about the early presence of human brain organization has recently been challenged by Dean Falk of the State University of New York at Albany. She believes, from her studies of specimens in South Africa and in Kenya, that brain organization in australopithecines was still basically apelike and that humanlike brain organization came about only in *Homo*.

Not for the first time do we have two experts in the field coming to opposite conclusions based on the same evidence. I am not a paleoneurologist, and never hope to be. Nevertheless, I do know that the material with which these people work—natural brain casts or rubber impressions of the inside of a cranium—is extremely difficult to interpret. The signals on the surface of brain casts are often faint, and false signals are common. As a result, the dispute between Ralph and Dean has been continuing in the pages of scientific journals for almost a decade. I once tried to bring Ralph and Dean together, while both were working in the Nairobi museum, in the hope that they could work out their differences, but with no success. Recently, though, Dean Falk's position has begun to receive support from other neurologists, including Harry Jerison.

It would make sense, and be consistent with other evidence, if a humanlike brain organization appeared with the origin of the genus *Homo*. So many other things changed at this point in our evolutionary history, associated with a major adaptive shift in the direction of hunting and gathering. Indeed, I would be surprised if a rudimentary spoken language had not been part of this *Homo* package. My guess is that it was absent in australopithecines, and that their mode of vocal communication was much like what we find in modern great apes.

We can turn to other aspects of fossil anatomy for clues to the emergence of language, namely, the vocal apparatus itself: the larynx, the pharynx, the tongue, and the lips. In the basic mammalian pattern the larynx is high in the neck, a position that has two consequences. First, the larynx locks into the nasopharynx —the air space near the "back door" of the nasal cavity—allowing the animal to breath and drink at the same time. Second, the range of sounds the animal can make is limited, because the pharyngeal cavity—the sound box—is necessarily small. Vocalization therefore depends mostly on the shape of the oral cavity and lips, which modify the sounds produced in the larynx.

In humans, the structure is very different and is unique in the animal world. The larynx is much lower in the neck, so humans cannot drink and breath simultaneously without choking. We also are much more vulnerable to choking on food while we are swallowing it. These are clear, negative results of the anatomical shift, so we can assume that the positive result is very striking indeed. And it is. The lower position of the larynx creates a much larger pharyngeal space above the vocal cords, allowing for a far greater range of sound modification. "The expanded pharynx is the key to our ability to produce fully articulate speech," says Jeffrey Laitman, of Mount Sinai Hospital Medical School, in New York.

Laitman was led to this question through his interest in the development of the human vocal tract during infancy. He found that human infants effectively recapitulate a segment of our evolutionary history. Babies are born with the larynx in the typical mammalian position, high in the neck, and can simultaneously breathe and drink, as they must during nursing. When the baby is about a year and a half, the larynx begins to migrate down the neck, reaching the adult position by about fourteen years. With this migration of the larynx comes an ever greater ability for sound production, as parents are only too well aware.

Laitman's work not only is fascinating in its own right, but it also gives us, potentially, a way to look for language in the fossil record: a high larynx in a human ancestor would imply apelike language ability, a low larynx, humanlike ability. The problem, of course, is that much of the structure of the vocal tract is composed of cartilage, which almost always decays during fossilization. All is not lost, however, as Laitman explains. "During our investigations, my colleagues and I noticed that the shape of the bottom of the skull, or basicranium, is related to the position of the larynx. This is not surprising, since the basicranium serves as the roof of the upper respiratory tract." Very simply, in the basic mammalian pattern the bottom of the cranium—the roof of the respiratory tract—is essentially flat. In humans it is distinctly arched. Here, then, is a signal that could be sought in the fossil record: the shape of the bottom of the crania of human ancestors.

"The pattern we see is very interesting," says Laitman. "First, in all the australopithecines I've examined, you see a typical apelike basicranium. This indicates to me that it would have been impossible for them to produce some of the universal vowel sounds that characterize human speech patterns. Second, the earliest time in the fossil record that you find a fully flexed basicranium is about 300,000 to 400,000 years ago, in what people call archaic *Homo sapiens*." Once again we have evidence that the australopithecines lacked something we regard as human. The results also indicate that archaic sapiens had a modern human larynx. This forces us to focus on what it was that changed with the emergence of fully modern humans, about 100,000 years ago.

We have to know, first, what happened in the middle of this sequence, between the australopithecines and the archaic sapiens, because it may indicate when an expansion of vocalization began. Laitman looked at cranium 3733, one of the earliest *Homo erectus* crania we have, and noted that basicranial flexion had already begun. "This individual would have had the ability to make a wide range of sounds. You could say that human spoken language had already started to evolve." The 3733 individual died about 1.6 million years ago, the same time that the Turkana boy lived, but on the opposite side of the lake.

Even though I've urged him to be specific, Laitman is not yet able to say precisely what kinds of sounds 3733 and his fellow early *Homo erectus* individuals could have made. "We haven't done the necessary computer modeling," he explains. "Probably they lacked certain vowel sounds, such as in *boot, father,* and *feet.*" What is clear, however, is that the larynx was no longer high up in the neck. Laitman guesses that in adult *erectus* individuals the larynx was in an intermediate position, between the ape configuration and what we have in modern humans, equivalent to the position in six-year-old modern humans.

"Yes, they almost certainly could have choked while drinking or eating," Laitman says of early *erectus*. "And you have to take that as a significant sign." The advantage of having the larynx in this intermediate position must have been considerable to outweigh the disadvantage of the dangers of choking. That advantage was surely a partially developed linguistic skill. I am not surprised.

If early *Homo erectus* had at least a rudimentary spoken language, what of *Homo habilis*, the earliest known member of the *Homo* lineage? Alas, we can't say. None of the *habilis* crania we have is sufficiently intact. If I had to guess I'd say that when we

263

do find an intact cranium of the very earliest *Homo*, we shall see the beginnings of the flexion in the base, the beginnings of the lowering of the larynx, the beginnings of spoken language.

The two areas of the fossil record that can tell us anything at all about language capabilities in our ancestors are therefore consistent with each other. Both indicate an early development of spoken language, starting almost certainly with the beginnings of the genus *Homo*. But the trajectory of that development, and its final elaboration, are less clear. We need to turn elsewhere in the record, to the products of our ancestors' minds: not their words, but the things they made.

A decade and a half ago Glynn Isaac was asked to make a presentation at a major congress on language origins, organized by the New York Academy of Sciences. "Asking an archeologist to discuss language is rather like asking a mole to describe life in the treetops," he said. "The earthy materials with which archeologists deal contain no direct traces of the phenomena that figure so largely in a technical consideration of the nature of language. There are no petrified phonemes, no fossil grammars. The oldest actual relics of language that archeologists can put their hands on are no older than the first invention of writing systems, five or six thousand years ago. And yet the intricate physiological basis of language makes it clear that this human ability has deep roots, roots that may extend as far back in time, or further, than the documented beginnings of tool making, some two and a half million years ago."

Inventive as ever, Glynn pondered the record with which he was most familiar—the stone-tool record—and sought ways in which it might pertain to the issue of language. He decided to study the changing complexity of stone-tool assemblages through time. Between about two million and fifty thousand years ago, the various elements of the assemblages became ever more numerous and ever more sharply defined, with major "improvements" occurring abruptly about 1.6 million years ago and some 250,000 years ago. These dates coincide with the appearance of, first, *Homo erectus* and then archaic sapiens. As human history progressed, stone-tool makers became better at making artifacts, turning out the different items to more standard form. In other words, choppers began to look more consistently like choppers, scrapers, more consistently like scrapers, and so on. This apparent improvement in stone-tool making has often been taken to indicate that our ancestors became more skilled in tool manufacture and were able to do more jobs with their tools, but Glynn questioned this assumption. "It's not necessarily true that the increase in complexity reflects an increase in the number of tasks performed with stone tools, nor are the fancy tools necessarily more efficient in an engineering sense," he remarked. "This is a point that has seldom been recognized."

If there is no real increase in efficiency, no broadening of tasks that could be performed with the more carefully elaborated stone artifacts, what does this pattern imply? Why should tool makers take care to produce artifacts to a more standard shape? "My intuition is that we see in the stone tools the reflection of changes that were affecting culture as a whole," Glynn said. "Probably more of all behavior, often but not always including tool-making behavior, involved complex rule systems. In communications, this presumably meant a more elaborate syntax and an extended vocabulary; in social relations, perhaps larger numbers of defined categories, obligations, and prescriptions; in subsistence, increasing bodies of communicable knowhow."

In other words, suggested Glynn, the imposition of order we see emerging in the stone artifacts throughout the archeological record is a cultural echo of what was happening in the rest of society. The social and economic contract that is at the heart of the hunting-and-gathering life demands of individuals an understanding of their roles, their place in the community, the behavior that is expected of them. Among foraging people in the modern world, elaborate kinship systems define an individual's relations to his fellows, to members of other groups, to his ancestors, even to gods. Such systems often dictate who may marry whom, who must share food with whom, and who may live where. Order is imposed on society by rules that prescribe acceptable behavior.

Imposition of order is a human obsession, a form of behavior that demands a sophisticated spoken language for its fullest elaboration. Granted, one can argue that birds impose order on their world when they construct elaborate nests, always to a prescribed design. But the important defining characteristic of humans is that the end products of its drive for order are highly individualistic to different societies. The arbitrariness is an element of human-imposed order, whereas the bird will always construct its nest in the same way. The obsession with ordering the world must have evolved during our history and without doubt is paralleled by the evolution of language. Without language, the arbitrariness of human-imposed order would be impossible.

I like the way that Glynn's hypothesis weaves together technology, language, and culture, producing a complex fabric that we recognize in ourselves and in our society today. Can we weave an esthetic sense into this fabric too? There is no doubt in my mind that the crafting of sculptures and painted images in the Ice Age involved esthetics, an eye for pleasing form. But we are dealing with modern humans, people like us. With our earlier forebears, it is difficult to call. Nevertheless, some of the hand axes I've seen from Africa and Eurasia are truly exquisite in form, the work of great care, patience, and, possibly, pride. Quite simply, they are better made than they need be for the utilitarian purpose of heavy-duty slicing and chopping. So I suggest that a sense of esthetics began to glimmer in *Homo erectus* and was enhanced in archaic sapiens, including Neanderthals. Both the imposition of arbitrary order and an emergent esthetics must have contributed to the shaping of our ancestors' worlds, and both must have required a degree of spoken language.

The later part of the archeological record, particularly from the Upper Paleolithic onward, shows accelerating change, of escalating innovation. The numbers of tool types increased, and many of them were delicate in ways not previously seen. Underlying the changes of this period almost certainly is a continuation of the process Glynn identified earlier in the record: the fabric of culture becoming ever more finely woven. In addition, there is real innovation here, the product of technical intelligence, not just social rules for the imposition of order.

The complexity evident among human groups in this late part of the record was surely accompanied by a well-developed spoken language. Glynn's real contribution—the mole describing life in the treetops—relates to the earlier part of the record, in which he identified a steadily augmented cultural milieu, the engine of which is language.

In the absence of direct evidence of spoken language, what other kinds of evidence can we seek? Some anthropologists argue that evidence of abstraction would be sufficient. A mind without language is effectively locked into the mental world it inhabits, because words and the reflective thoughts they allow are the only tools for exploring the edges of that world, seeking transcendance of it. Words can create experiences that have not happened: they are the fuel of imagination, of conceptualization. Visual images, it seems to me, are a unique product of such conceptualization, the evidence of abstraction that we seek.

In spite of long, patient encouragement, apes have so far failed to paint images that an objective observer can accept as representational. This doesn't surprise me. The realization that lines drawn on a surface or carved into bone or ivory may be

representational was an event in human history of great intellectual magnitude, equal to many of the greatest of scientific discoveries. It was the product of a mental world explored beyond
its edges by means of language. Even more difficult to imagine in
the absence of language is the conceptualization of symbolic
images, whether representational or abstract. The simple image
of a cross, for instance, or a shepherd with a lamb, carries powerful connotations in Western culture; they are innocent symbols
of an entire religious mythology. I suspect that some of the
representational images we find in the archeological record are
part of a mythology, for without language, there can be no mythology.

When we find representational and abstract images of the
sort that begin to appear about 30,000 years ago in Africa and in
Europe, we surely are dealing with people endowed with a fully
modern, articulate, spoken language. I am much less certain,
however, that the absence of such images earlier in the record
necessarily implies an absence of anything approaching modern
language. And I would not go as far as the New York University
anthropologist Randall White, who says that earlier than about
100,000 years ago there was "a total absence of anything that
modern humans would recognize as language." White bases his
position on the dramatic changes he sees with the beginning of
the Upper Paleolithic, like an increase in the size of social
groups, evidence of trade, unprecedented technological innovation, and, of course, art.

It is indeed striking that image making comes into the record
suddenly and recently, about thirty thousand years ago in both
Africa and Europe. (The best-known cave paintings of Europe
are later, beginning about twenty thousand years ago.) Earlier
than this there are merely scattered indications of some kinds of
symbolic behavior: a simply engraved ox rib from the 300,000-
year-old site of Pêch de l'Azé, in southwest France, or a piece of
sharpened ocher in a coastal shelter, constructed some 250,000

years ago near Nice, in southern France. Beyond this, there is very little.

The engraving on the ox rib is a series of festooned double arches, a pattern reminiscent of engravings to be found in the same area from thirty thousand years ago onward. Does this point to a continuity of tradition, across a vast tract of time, empty of other examples? I doubt it. More likely, the pattern represents something basic in the human psyche. The piece of ocher, sharpened as if it had been used for coloring, is redolent of ritual. But the virtual emptiness of the record before thirty thousand years ago is puzzling.

If, as seems likely, fully modern humans evolved about 100,000 years ago, why is it that we don't find evidence of artistic and symbolic expression until some seventy thousand years later? It is possible, but unlikely, that before thirty thousand years ago all manifestations of symbolic behavior were on materials, like sand and bark, that did not survive. And as paintings can survive on rock shelter walls and deep in caves for thirty thousand years, they certainly could have survived forty thousand or fifty thousand or even 100,000 years. The signal in the record thirty thousand years ago appears to be real, whatever it means.

Symbolism can be encapsulated in things other than images, of course. Ritual burial is the most pertinent example, often associated with Neanderthals: the body of a hunter, laid out in a grave in the cave of La Chapelle-aux-Saints in France, forty thousand years ago, together with a bison leg, other animal bones, and some flint tools, for example; a woman buried in an exaggerated fetal position in the cave of La Ferrassie, in France, one of half a dozen graves in the site. There are many such examples described in the literature.

Perhaps the most famous is the old man of Shanidar, in the Zagros Mountains of modern Iraq. Sixty thousand years ago he died and is said to have been laid out on a bed of plant material, surrounded by spring flowers: yarrow, cornflowers, St. Barnaby's thistle, groundsel, grape hyacinths, woody horsetail, and a kind of mallow. The flowers of white, yellow, red, blue, and purple also have medicinal value. The old man of Shanidar has therefore been said to be a shaman, or medicine man, and his burial a ceremony befitting such an important member of the group.

Concern for the dead expressed in these ways indicates a well-developed language and a well-developed consciousness. Self-awareness and death-awareness flow together. Are we seeing here, then, a continuation of the language development we infer for earlier periods in the record? I believe we are. Just recently some of the claims for evidence of Neanderthal burial have been challenged, particularly by Robert Gargett, of the University of California at Berkeley.

Gargett suggests that virtually all the putative burials can be explained as natural burials—the collapse of cave roofs on occupants or abandoned bodies, for instance—devoid of ritual. He is probably correct for some instances, where evidence may have been overinterpreted. But there are too many cases where chance would have to be invoked as an explanation of associations of bodies and stone tools, of alignments of bodies, and so on. The evidence is still convincing that Neanderthals, and probably other archaic sapiens, occasionally buried their dead with a degree of ritual that we recognize as human.

In this context the language capabilities of Neanderthals is pertinent. Unfortunately, there is no consensus among the experts. "Poor *Homo sapiens neanderthalensis*," laments Ralph Holloway. "Surely no other ethnic group has had so many nasty slurs and insults thrown at it than our distant cousins of some forty thousand to fifty thousand years ago . . . The final blow is the attitude, based on computer decisions and a lack of art work, that poor Neanderthals were also mute, or at least babbling away with a highly restricted set of phonemes." For Ralph, the paleoneurological evidence shows two things: the Neanderthal brain is fully *Homo*, with no significant differences from that of modern humans; and "Neanderthals did have language."

The "final blow" to which Ralph refers came from Philip Lieberman, a linguist at Brown University. Based on a study of the basicranial anatomy of the old man of La Chapelle-aux-Saints, Lieberman and his colleague Edmund Crelin concluded that Neanderthal speech was severely limited. "Classic Neanderthal hominids appear to be deficient with respect to their linguistic and cognitive ability," Lieberman said recently. "At minimum their speech communications would have been nasalized and more susceptible to perceptual errors—they probably communicated vocally at extremely slow rates and were unable to comprehend complex sentences."

The basicranium of the old man of La Chapelle is no more flexed than what we see in 3733, a *Homo erectus* from 1.5 million years earlier in our ancestry. Does this mean that in Neanderthals, the larynx was in the same position in the neck as in early *erectus*; that Neanderthals' language was no more developed than it had been 1.5 million years earlier; indeed, that Neanderthals' language abilities had regressed from what had been achieved in other archaic sapiens?

The conclusions of Lieberman and his colleagues suggested that language deficiency played a major role in the disappearance of Neanderthals. The issue, it turns out, is not at all clear-cut. For one thing, the old man of La Chapelle is a badly deformed skeleton in many ways, and it is quite possible that the degree of flexion in the basicranium was more pronounced than it appears. Nevertheless, according to Jeffrey Laitman, who has worked with Lieberman and Crelin, basicranial flexion in some Neanderthals is not as modern as in most archaic sapiens individuals. Other Neanderthals do fall within the range of what would be called modern. "I think we're looking at something very complicated," cautions Laitman.

I asked Laitman to "grade" the Neanderthals in relation to their assumed vocal tract structure, using an arbitrary scale of 1 to 10, where 1 represents the ape grade and 10 the modern human grade. He said that the old man of La Chapelle would be about 5, and other Neanderthals would be in the 7-to-8 range. Remember, the earliest archaic sapiens, from about 300,000 years ago, clock in at a fully human 10 on this scale. This means that Neanderthals had regressed halfway back to the ape grade. In fact, they would have been about equal to 3733, the early *erectus* specimen, which Laitman says is about a 6 on this scale. I find it hard to conceive of such a regression in a function that seems to have been selected for so very strongly during human history.

The picture was made even more complicated—or confused —just recently with the remarkable discovery of a little bone from a sixty thousand-year-old Neanderthal in the Kebara Cave in Mount Carmel, Israel. The partial skeleton was discovered in 1983, by a joint French-Israeli expedition, and has provided some interesting insights into Neanderthal anatomy. Most significant of these is the hyoid bone, a tiny U-shaped bone that anchors muscles connected to the jaw, larynx, and tongue. Because of its central position in the vocal apparatus, the hyoid is vital to voice production. The Kebara hyoid is the first to be recovered from an ancestral human, the first glimpse of this crucial piece of anatomy.

"We conclude that the morphological basis for human speech capabilities appears to have been fully developed," said the excavators, led by Baruch Arensburg of Tel Aviv University and Bernard Vandermeersch of the University of Bordeaux. The anatomy of the Kebara hyoid was indistinguishable from that in modern humans. "The assumed speech limitations of the Neanderthals, which have hitherto been based primarily on studies of basicranial morphology, would seem to require revision," they added with studied understatement.

Philip Lieberman was unconvinced by the evidence. "There's no basis for comparison, since we have no hyoids from earlier hominids," he told a reporter for *Science News*. "At this point, the Kebara hyoid doesn't tell us anything about the evolution of speech and language." Jeffrey Laitman is equally cautious. "The anatomy of the hyoid doesn't give us enough information to reconstruct the structure of the vocal tract," he says.

Clearly, my colleagues are not yet close to agreement on this issue. It seems to me that there is a complicating factor that has not yet been fully assimilated. Earlier, I described the unusual anatomy of the Neanderthal face: midfacial protrusion, as if it had been pulled out by the nose. This configuration produces large air spaces in the upper respiratory tract, which have been interpreted as structures for warming inhaled frigid air and for condensing water vapor in exhaled air. The Neanderthals were quintessentially cold-adapted people, and such functions could have been an important aspect of that adaptation.

It is surely possible that the unusual structure of the upper respiratory tract affected the shape of the basicranium, perhaps without altering the position of the larynx. No one can be certain, but I think this is more likely than the alternative: that the changes in structure of the upper respiratory tract due to cold adaptation severely compromised the Neanderthals' ability to produce a wide range of sounds, thus reducing their language capabilities. I find it hard to imagine that Neanderthals, with brains slightly larger than the average modern human brain, were language imbeciles. Their technology was as developed as that of other archaic sapiens, and more so in most cases. And their expression of self-awareness, through ritual burial, was as developed. Neanderthals must have been as well endowed linguistically as any other archaic sapiens population. But not as well endowed as modern humans.

The major event in the origin of modern humans very likely was the final acquisition of a fully articulate spoken language. The evidence we have seen supports that notion. To suggest otherwise—to imagine a human species that is equipped with a language like ours and yet is not us—is to me impossible. I believe that the final evolutionary step was an incremental change, not a revolutionary punctuation. The crucial advance may not have been in the capability or rate of sound production, however. Instead, it may have been in sound perception, the mental decoding machinery. Language abilities, after all, have both production and perception components, which much evolve more or less in concert. Human speech is built from fifty sounds, compared with a dozen in other higher primates. That fourfold advance is the result of a modified vocal tract, some aspects of which we can detect in the fossil record. But the advance in breadth and rate of communication that accompanied those changes in the vocal tract is far greater than fourfold: it is an infinite advance over our primate cousins. This, almost certainly, is the result of the remodeling of mental machinery within the brain, signs of which probably never impress themselves directly on the fossil record.

With the origin of modern humans there was seeded in the world a great linguistic blossoming: from a single language more and more developed, evolving in ways analogous to the evolution of species but at a much faster rate. After perhaps 100,000 years, some five thousand languages existed, the number documented in recent historic times. Five thousand languages, each rooted through complex evolutionary relationship to an original mother tongue. Five thousand languages, each belonging to one of a dozen or so language families, shadows of that deeper relationship. Five thousand languages, each an expression of an ability that unifies all of humankind.

And yet, ironically, the cognitive ability that unifies all of *Homo sapiens* also fragments *Homo sapiens*. For five thousand languages means five thousand cultures, each a social and spiritual milieu that differentiates and, all too often, separates one from another. At birth each of us is capable of speaking any one of those five thousand languages—indeed, any one of an infinity of human languages—but under natural circumstances we learn just one. As the Princeton anthropologist Clifford Geertz puts it: "One of the most significant facts about us may finally be that we all begin with the natural equipment to live a thousand kinds of life but end in the end having lived but one." The irony is that, through language, individuals come to understand themselves, their society, and their culture, all foreign to individuals from other cultures. The medium for understanding can be a barrier to understanding, a result of the power, not the limitation, of language.

Throughout human history, humans increasingly created their own environment, that of culture. This unique artifact is so dominant a force in our lives that we come ultimately to depend on it. As Geertz so eloquently puts it: "A cultureless human being would probably turn out to be not an intrinsically talented though unfulfilled ape, but a wholly mindless and consequently unworkable monstrosity. Like the cabbage it so much resembles, the *Homo sapiens* brain, having arisen within the framework of culture, would not be viable outside of it." No other living creature is so constrained or so liberated by its heritage.

This statement looks very much like a different version of Thomas Huxley's assessment of the import of human language: that we "stand raised upon it as on a mountaintop, far above the level of his humble fellows." In a sense it is true. Human language, and all that flows from it indeed marks *Homo sapiens* as an especially talented species. And we can very probably justify a claim to being more than just especially talented. Our sense of morality, ethics, and transcendental vision is unique in today's world.

What, then, of the "vastness of the gulf between . . . man and the brutes" that Huxley identified? I have suggested that recent work on cognitive capacities—particularly language capacities—of higher primates to some extent closed the gulf at the nonhuman end of things. I argue, too, that what we learn from human prehistory serves to close it still further. We may feel special in today's world, and in many ways we are, but we are merely the end product of an evolutionary lineage that joins us through unbroken links to the rest of the natural world.

If we learned from fossil and archeological evidence that language ability had emerged only with the origin of modern humans, then we might claim with justification that *Homo sapiens* was indeed separated from "the brutes" in some significant way. Instead, we learn that language ability—and with it, human culture and consciousness—emerged gradually through our history. Each of the *Homo* species antecedent to *Homo sapiens* had a measure of humankind about it. Each had a frisson of humanness about it, not just in stature and deportment, but in the way the mind worked. The beacon of humanness burned ever more brightly through time, until it illuminated the world with the glaring intensity that we now experience.

If, by some freak of nature, *Homo erectus* and *Homo habilis* still existed, *Homo sapiens* would appear to be far less special than we like to think of ourselves. The gap between humankind and "the brutes" would be filled by *erectus* and *habilis*, and our connectedness with the rest of the natural world would be much more evident. These species—*erectus* and *habilis*—do not exist, of course, except as elements of an evolutionary record. For this reason, an understanding of the human fossil record is both poignant and instructive, for it reveals to us our true place in the world, and puts our undoubted specialness in historical perspective.

16

Murder in a Zoo

For five years Luit had been jockeying for leadership of the chimp colony at Burgers' Zoo, in Arnhem, the Netherlands. Intermediate in age between Yeroen and Nikkie, chief rivals for top chimp, Luit was a fine physical specimen, muscular, with a sleek, black coat. Nevertheless, Luit exploited his wit, not his physique, in his drive for leadership. Manipulating the balance of power first with Yeroen, then with Nikkie, sometimes playing off one against the other, Luit finally achieved the position of alpha male, and with it the favored access to the females in the colony. But success was soon followed by disaster, and this time brawn, not brain, prevailed: Yeroen and Nikkie joined forces and attacked Luit, brutally.

"I was working at home," remembers Frans de Waal, who has studied the Arnhem colony for many years. "It was a Saturday morning. The telephone rang, and the news was about as bad as it could be." Rapidly, and in great distress, de Waal's assistant described how she had just found Luit, barely conscious and covered in blood and torn flesh. Deeply gashed on his head, sides, hands, and feet, Luit appeared near death. The most gruesome of injuries: Yeroen and Nikkie had torn off both of Luit's testicles. "Do what you can for him," de Waal shouted into the phone. "I'll be there right away."

Emotions churned in his breast as he frantically cycled the short distance between his house and the zoo, emotions of despair and sadness. And accusation. "Yeroen was to blame," he kept thinking. Luit's condition was even worse than de Waal had feared, and despite three hours of surgery, Luit died, from a combination of stress and loss of blood. Even now, a decade after the incident, de Waal says, "I cannot look at Yeroen without seeing a murderer. Nikkie, ten years younger than Yeroen, was only a pawn in Yeroen's game."

What kind of game could end in murder? "Politics, the pursuit of power," explains de Waal. Two millennia ago, Aristotle labeled the human a *politikon zoön*, a political animal. "He could not know just how near the mark he was," says de Waal. "Our political activity seems to be part of an evolutionary heritage we share with our close relatives." Assassination as the resolution of power struggles is not unusual in the pages of human history, a last resort when the usual means of political manipulation fail. So it had been for Luit. "What my work at Arnhem has taught me," says de Waal, "is that the roots of politics are older than humanity."

Although many people are unaware of the fact, my father believed fervently that if we understood more about the behavior of the great apes—our closest living relatives—we would understand more about ourselves and about our evolutionary history. He devoted considerable energy to establishing what eventually became two of the most famous and influential of all primate field studies: Jane Goodall's, with chimpanzees in Tanzania, and Diane Fossey's, with gorillas in Rwanda. The information gained from studies of the social behavior of the African apes would complement what we could learn from the fossil record, my father argued. How right he was.

Not only have these and other primate studies given us concrete ways of thinking about the social structure of our ancestors (especially in relation to such factors as body size, ecology, and sexual dimorphism); they have given us insights into the primate mind. They have given us a glimpse of the evolution of consciousness, the phenomenon that we experience as introspection and self-awareness.

In recent years primatologists and psychologists have come to realize just how Byzantine is the primate mind. From the monitoring and manipulation of complex networks of relations, to the perpetration of clever tricks of deception, our primate cousins inhabit social and intellectual worlds more complex than we could have imagined. The title of a recent book on the subject says it all: *Machiavellian Intelligence*.

That the British psychologists Richard Byrne and Andrew Whiten, editors of *Machiavellian Intelligence*, considered such a title appropriate for a scholarly text on primate social expertise is indicative of the respect now being accorded to the nonhuman primate mind. We look at the subject here as a way to understand something of the human mind and, in particular, of the emergence of introspective consciousness during our evolutionary history.

More than any other subject, the nature of mind has troubled, tantalized, and eluded philosophers and psychologists alike. Operational definitions, such as "the ability to monitor your own mental states, and the corresponding capacity to use your experience to infer the experience of others," may be objectively accurate, but they don't capture the essence of what each of us *feels* "mind" to be. Mind is the source of the sense of self, the private world the self inhabits, a world that is sometimes shared with others; it is the font of hopes and fears, of good and evil; it

Louis and Mary Leakey display the palate of Zinjanthropus shortly after it was discovered, in 1959. (Des Bartlett)

Raymond Dart (right) examines the Taung skull in company with Phillip Tobias (left) and Fred Grine, during the "Ancestors" exhibit at the American Museum of Natural History, in 1984. (American Museum of Natural History)

P*art of the famous trail of footprints at Laetoli, Tanzania, which were made by early hominids,* 3.6 *million years ago. (Andrew Hill)*

M*ilford Wolpoff, chief spokesman for the multiregional hypothesis for the origin of modern humans. Wolpoff says: "You can't imagine one human population replacing another except through violence." (University of Michigan)*

Christopher Stringer, strong supporter of the Out of Africa hypothesis for the origin of modern humans. Stringer says: "Paleoanthropologists who ignore the increasing wealth of genetic data on human population relationships will do so at their peril." (The Natural History Museum, London)

Holly Smith, who searched for clues to life history factors of Homo erectus by studying the Turkana boy's teeth. Smith says: "It was fortunate for me that the Turkana boy died the way he did." (The University of Michigan)

*G*ordon Gallup, who developed a test for consciousness in nonhuman primates. Gallup says: "The first time we tried it with chimps it worked." (State University of New York, Albany)

*D*ean Falk studies the structure of fossil brains and concludes that the humanlike brain organization developed with Homo but not Australopithecus. (Dean Falk)

East Lake Turkana, from the air. The Koobi Fora camp can be seen as a collection of thatched huts (bandas) at the base of the Koobi Fora spit. (P. Kain/Sherma)

This Neanderthal skeleton was recently excavated at the Kebarra Cave, Israel, by a joint Israeli–French team. (O. Bar-Yosef and B. Vandermeersch)

a

b

In the power struggle between the dominant chimpanzees at Burgers' Zoo in Arnhem, Nikkie and Yeroen at one point formed a coalition. In *a*, we see Nikkie (left) becoming angry that his coalition partner (Yeroen, right) is sitting near to Nikkie's rival, Luit. Nikkie sits aggressively opposite the other two males. In *b*, Nikkie bluffs a threat, and Yeroen gets up and moves away. In *c*, his separation attempt now successful, Nikkie bluffs ostentatiously over Luit. (Frans de Waal)

c

Hadza women, in East Africa, collect tubers as part of a division of labor typical among foraging people. (James F. O'Connell)

Site 50, on the eastern shore of Lake Turkana, provided a means of testing hypotheses of the social and economic organization of Homo erectus. (Roger Lewin)

Olduvai Gorge, Tanzania, where Louis and Mary Leakey spent many years searching for early human fossils and artifacts. When Zinjanthropus was discovered here in 1959, it was the first australopithecine known outside of southern Africa. Mary established modern paleolithic archeology with her analyses of ancient stone tools (Oldowan and Acheulean assemblages). (University of California)

is the means of experiencing worlds beyond the tangible and the means of making the intangible real.

"Few questions have endured longer or traversed a more perplexing history than this, the problem of consciousness and its place in nature," says Julian Jaynes, a Princeton psychologist. This "problem" is at the heart of our striving to understand our humanness. And, unexpectedly, this is what many primate studies—including, for instance, the one about the circumstances surrounding Luit's death—are beginning to illuminate. The same light will elucidate something of the mind of *Homo habilis* and *Homo erectus* and, ultimately, *Homo sapiens.*

For René Descartes, three centuries ago, the mind and the body were entirely separate entities, a dualism that made a whole. "It was a vision of the self as a sort of immaterial ghost that owns and controls a body the way you own and control your car," observes the Tufts University philosopher Daniel Dennett. Philosophers have referred to this issue as the mind-body problem. "More recently, with the rejection of dualism and the rise of materialism—the idea that the mind just *is* the brain—we have gravitated to the view that the self must be a node or module in the brain, the Central Headquarters responsible for organizing and directing all the subsidiaries that keep life and limb together."

I take the materialist view that consciousness is the product of the brain's activity, not some gossamer attachment to the organ, as the noted neurologist Sir John Eccles recently suggested. "I am constrained to attribute the uniqueness of the Self or Soul to a supernatural spiritual creation," he wrote in his latest book, *Evolution of the Brain.* While rejecting external intervention of an Ecclesian nature, I am nevertheless sympathetic to the sentiments expressed in a recent essay by Colin McGinn, a philosopher at Rutgers University.

"How is it possible for conscious states to depend on brain states?" ponders McGinn. "How can Technicolor phenomenology arise from soggy gray matter? What makes the bodily organ we call the brain so radically different from other bodily organs, say the kidneys—the body parts without a trace of consciousness? How could the aggregation of millions of individually insentient neurons generate subjective awareness? . . . It strikes us as miraculous, eerie, or even faintly comic. Somehow, we feel, the water of the physical brain is turned into the wine of consciousness, but we draw a total blank on the nature of this conversion."

In saying that the consciousness is the product uniquely of the brain, McGinn is emphasizing Dennett's point: the mind *is* the brain. He is also saying that consciousness is of the brain and of no other organs, no other matter in the world. Now, I am aware that some scholars argue that nonbrain matter may also be conscious in a way, matter such as other organs in the body, plants, even rocks, and, at the extreme, fundamental particles of matter. This seems to me a philosophical exercise of little relevance to what we, as humans, actually experience. My concern is with that numinous sense of self, undeniable yet undefinable. I therefore enthusiastically support McGinn's sentiments, both on the source of consciousness and its frustratingly evanescent nature.

McGinn is not saying that, because we cannot explain the physical process by which nonsentient material generates sentience, the process must be miraculous, beyond physical laws. He is suggesting that, difficult though we find it to accept, there may be limits to human understanding about nature; the human brain may not be equipped ultimately to understand itself. "We find it taxing to conceive of the existence of a real property, under our noses as it were, which we are not built to grasp," he says; "a property that is responsible for phenomena that we observe in the most direct way possible [our own experience]."

Three great biological revolutions mark the history of life in the world. The first is the origin of life itself; the second, the origin of eukaryotic cells, cells with nuclei; the third, the origin of multicellular organisms. Each of these revolutions transformed the world in dramatic ways. It is not an exaggeration to add a fourth revolution: the origin of human consciousness. It represents a new dimension in biological experience.

Our question here, as it was with the origin of language, is whether consciousness in the way we experience it arose rapidly and recently in our history. Or was it built gradually, resting on cognitive foundations in our apelike ancestors? In our quest for the roots of human consciousness we will explore the social worlds of nonhuman primates; we will ask why primates seem to be more intelligent than they need be; we will search for signs of a sense of self in these animals, including their penchant for deceiving one another; we will look at the meager clues of consciousness in the prehistoric record; and we will address the emergence of mythology and religion.

The first question to ask is not so much the what of consciousness as the why. Why should there even be the phenomenon of human consciousness? The capacity for introspection experienced by humans may be an epiphenomenon of a large, complex brain, the byproduct of other neural functions, "the whine of neural gears, the clicking of neural circuitry," as the Cambridge University psychologist Horace Barlow once put it. In my approach to the human species, however, it is necessary to view consciousness as we view other aspects of ourselves: the direct product of natural selection. In that case, we have to ask what selective benefit consciousness conferred on our ancestors and on us. I will take what may seem a circuitous route in answering this question. The route begins with asking why higher primates appear to be more intelligent than they "need" to be.

. . .

At the Language Research Center of Georgia State University, in Atlanta, there is a monkey that, armed with a miniature joystick, can anticipate the complex movement of an object on a computer screen and ultimately "capture" the object. Even for a human, the task is not easy. It requires concentration on, and prediction of, the object's likely trajectories, as well as fine manipulation of the joystick. And yet the monkey is not specially trained for the task, not especially talented. Any of the monkeys at the center can do it, once they are familiar with the system. In another part of the center are to be found chimpanzees that can accomplish even more demanding intellectual problems, often ones that require the ability to see three or four moves ahead in a sequential puzzle. Analytical skills, reasoning, and foresight are called for. Again, these animals are not specially trained or especially talented. They are displaying the natural talents of higher primates.

This presents us with a conundrum, for what is natural about the talents I've just described? Psychologists who study the cognitive abilities of monkeys and apes in laboratory situations agree that the animals seem to be profoundly smarter than their natural needs demand. "During two months that I spent watching gorillas in the Virunga Mountains of Rwanda," observes another Cambridge University psychologist, Nicholas Humphrey, "I could not help being struck by the fact that of all the animals in the forest the gorillas seemed to lead much the simplest existence—food abundant and easy to harvest (provided they *knew* where to find it), few if any predators (provided they *knew* how to avoid them) . . . little to do (and little done) but eat, sleep, and play." The cognitive skills displayed by higher primates in the laboratory seem to outstrip by far the practical demands of their natural worlds. Has natural selection been profligate in making them smarter than they really need to be?

A few years ago Nick Humphrey visited Kenya, and we went to Koobi Fora, where we talked about his ideas. What, I wanted to know, did these observations have to do with the evolution of the human mind? "The same things you can say about the daily lives of gorillas—that the world of practical affairs seems not very demanding intellectually—you can say about humans," he replied. "Studies on hunter-gatherer societies show that the demands of their daily lives are not great. Hunting techniques do not greatly outstrip those of other social carnivores. And gathering strategies are of the same order as you might find in, say, chimpanzees or baboons."

I acknowledged this and wondered what it was in evolutionary history that enabled the human brain to create a Mozart symphony or Einstein's theory of relativity. "The answer," said Nick, "is social life. Primates lead complex social lives. That's what makes them—and has made us—so intelligent." I must admit to having been pretty skeptical about Nick's suggestion during that trip to Koobi Fora. The notion that the exigencies of social interaction, such as building alliances and outwitting potential rivals, may have been responsible for honing human intellect seemed somehow insubstantive. Perhaps it is because the social nexus is so natural a part of human existence that it becomes, in a way, invisible to our thought. A decade of research on nonhuman primates has, however, had the effect of making the social nexus not only visible but also sharply focused. Nick's hypothesis now looks very powerful indeed.

For a long time anthropologists accepted the idea that technology, not social interaction, was the driving force behind the evolution of the human intellect. Given that our physical world is dominated by the fruits of clever invention, it is not surprising that we are impressed by human technological skills. And it is natural that such skills should be thought of as the direct products of natural selection. But, as Harry Jerison has argued, it seems more likely that these skills are the byproduct of an intellect sharpened by other forces in natural selection. "Building a better reality" was how Jerison described it. As we bring this into closer focus, we can see that for the higher primates, the most important—and most intellectually challenging—components in an individual's reality are other individuals.

For a baboon or a chimpanzee, a certain amount of intellectual capacity and memory is required for exploiting food resources year round. They need a kind of mental map of their range. They need to know, and be able to recognize, when certain trees will be in fruit, when certain tubers are ready, when certain waterholes will be full. But there is a degree of predictability about it all, a pattern to follow. By contrast, other individuals in one's troop may be anything but predictable, particularly in their response to one's own behavior. A richly laden fruit tree may be difficult to find, but once located it doesn't suddenly disappear or its fruit become inedible. Such Alice-in-Wonderland transformations in a fruit tree would be the equivalent to the range of responses one individual in a troop may face when meeting another, a rival, for instance.

Compare gorillas and zebras, on the face of it an odd pair to contemplate together. Both species inhabit similar ecological worlds, in that both feed on evenly distributed and widely abundant food resources (leaves in mountain forests for gorillas, grasses on open plains for zebras). And in both species females leave their natal groups to live with a single dominant male and a group of unrelated females. Since their ecological worlds and social structures have the same framework, one might suppose their mental capacities would be similar too. Not so. Relatively speaking, gorillas are about four times brainier than zebras, a difference that correlates with a much more complex and demanding social life than zebras.

"Like chess, a social interaction is typically a *transaction* between social partners," says Nick Humphrey. "It asks for a level of intelligence that is unparalleled in any other sphere of living. There may be, of course, strong and weak players—yet, as master or novice, we and most other members of complex primate societies have been in this game since we were babies." Nick has been developing this line of argument since the early 1970s, and his ideas served to crystallize similar thoughts among other researchers. By now, the notion of social intelligence—or, rather, the acute intellectual demands of complex social life—has become the leading paradigm among anthropologists.

A recent review of primate studies, by Dorothy Cheney, Robert Seyfarth, and Barbara Smuts, confirms the ascendancy of the paradigm. "Nonhuman primate tool use, which has received considerable attention because of its relevance to human evolution, is striking in part because it is relatively rare," they note. "By comparison, primatologists repeatedly emphasize the ability of the subjects to use other individuals as 'social tools' to achieve particular results." They conclude that, "among nonhuman primates, sophisticated cognitive abilities are most evident during social interactions."

What is it that makes primate social life so complex that it demands "sophisticated cognitive abilities"? In a word, the principal element is *alliances*. As in all animal groups, the ultimate driving factor in individual behavior is reproductive success. In anthropomorphic terms, females strive to raise to maturity as many offspring as they can; males strive to father as many offspring as they can. For females, reproductive success is achieved through being able to care for and protect offspring; for males, reproductive success depends on having as many mating opportunities as possible. For both males and females, the goals are made easier if they can rely on the support of others, friends and relations. A great deal of primate life is therefore spent in nurturing such alliances for oneself and in assessing the alliances of one's rivals.

Consider the affair between Alex and Thalia, young mature baboons, members of a troop that live near Eburru Cliffs, 100 miles northwest of Nairobi, on the floor of the Great Rift Valley. Barbara Smuts, a primatologist at the University of Michigan, studied the social life of this troop over several years, paying particular attention to the establishment of what she called "friendships," long-term alliances between males and females. Alex had entered the troop recently and needed to establish a strong alliance with a female. He chose Thalia.

"It was like watching two novices in a singles bar," reports Barbara. "Alex stared at Thalia until she turned and almost caught him looking at her. He glanced away immediately, and then she stared at him until his head began to turn toward her. She suddenly became engrossed in grooming her toes. But as soon as Alex looked away, her gaze returned to him. They went on like this for more than fifteen minutes, always with split-second timing. Finally, Alex managed to catch Thalia looking at him. He made the friendly eyes-narrowed, ears-back face and smacked his lips together rhythmically. Thalia froze, and for a second she looked into his eyes. Alex approached, and Thalia, still nervous, groomed him. Soon she calmed down, and I found them together on the cliffs the next morning." Six years later, on a subsequent visit to Eburru Cliffs, Barbara found that the friendship formed in that initial tentative manner had held fast.

"Because baboon friendships are embedded in a network of friendly and antagonistic relationships, they inevitably have repercussions extending beyond the pair," explains Barbara. For instance, she once observed Cyclops with some meat, part of an antelope he had caught. Triton, the prime adult male of the troop, spotted the prize and began challenging Cyclops for it. "Cyclops grew tense and seemed about to abandon the prey," Barbara recounts. "Then Cyclops's friend Phoebe appeared with her infant, Phyllis. Phyllis wandered over to Cyclops. He immediately grabbed her close and threatened Triton away from the prey."

The move was a smart one, because if Triton had continued to threaten Cyclops, he would in effect be threatening Phyllis too, Phoebe's infant. As a result, "Triton risked being mobbed by Phoebe and her relatives and friends," explains Barbara. Against such odds, the prudent thing for Triton to do was back down, which he promptly did. "Thus, friendship involves costs as well as benefits because it makes the participants vulnerable to social manipulation or redirected aggression by others."

Networks of alliances hold primate troops together and govern individuals' interactions. Dorothy Cheney and Robert Seyfarth describe another example, this time in vervet monkeys, which they have studied in Amboseli National Park, in Kenya. "In a typical encounter, one female, Newton, may lunge at another, Tycho, while competing for a fruit. As Tycho moves off, Newton's sister Charing Cross runs up to aid in the chase. In the meantime, Wormwood Scrubs, another of Newton's sisters, runs over to Tycho's sister Holborn, who is feeding sixty feet away, and hits her on the head." What to the casual observer may look like an outburst of random aggression among a group of irascible individuals is in fact the playing out of conflict over ever-widening networks of alliances, both of blood and friendship.

"Hostility between two animals often expands to include whole families, so not only must monkeys predict one another's behavior, but they must assess one another's relationship," explain Dorothy and Robert. "A monkey confronted with all this nonrandom turmoil cannot be content with learning simply who's dominant or subordinate to herself; she must also know who's allied to whom and who's likely to aid an opponent." Through a series of ingenious experiments, in which they played recordings of specific individuals' calls, Dorothy and Robert were able to determine by the other monkeys' reactions that the animals know very clearly the patterns of relationship and alliance networks in their troop.

If alliance networks were permanent structures within a troop, it would be difficult enough for individuals to cope with their intricate connections. But they are by no means permanent. Always looking to their own best interests, and to the interests of their closest relatives, individuals may sometimes find it advantageous to break existing alliances and form new ones, perhaps

even with previous rivals. Troop members therefore find themselves in the midst of changing patterns of alliances, demanding
yet keener social intelligence to be able to play the changing
game of social chess.

Luit, Yeroen, and Nikkie were pieces in such a game of social
chess as their strategies propelled them toward that final, fatal
attack, in September 1980. Frans de Waal, who watched the
game, recorded it in detail.

In the beginning, in 1975, Yeroen, the eldest of the three,
was the unquestioned alpha male. Luit and Nikkie routinely sub-
mitted to Yeroen, and he enjoyed the allegiance of all the females in the troop. Then, in the summer of 1976, Luit stopped
showing submission to Yeroen, and began to challenge him with
noisy displays. The much younger Nikkie joined in support of
Luit, but only when Luit was confronting the females, Yeroen's
supporters. After a few months, Luit's challenges to Yeroen's position prevailed, and he became the dominant male in the troop.

During the first year of the new era, both Luit and Nikkie
solicited support from Yeroen, the fallen leader. It was as if Luit
knew he would be better off if Yeroen were a friend rather than
an enemy. Nikkie seemed to be seeking an alliance with Yeroen,
perhaps to try to overthrow Luit. In any case, at the end of the
year, Nikkie got what he wanted when Yeroen and he formed an
alliance against Luit. By this time Luit had gained the loyalty of
all the females in the troop, a state of affairs that was to bring his
downfall, suggests de Waal.

"Luit's fate is reminiscent of the balance-of-power paradox
'strength is weakness,'" de Waal explains. "This means that the
strongest of the three competing parties almost automatically
elicits cooperation against himself, because the weaker parties
gain more by joining together and sharing the payoffs than by
joining the strongest party, who will monopolize the payoffs."
As it was, with Yeroen's support, Nikkie became the alpha male.
By contrast with Luit's previous position, Nikkie's status de-

pended completely on the alliance with Yeroen. Although he wasn't the prime male, Yeroen still enjoyed considerable reproductive success, his access to females being tolerated by Nikkie as part of the deal between them.

Nikkie managed to consolidate his position over the next year, with Yeroen playing politics on both sides. He had been able to gain access to estrous females, sometimes by enlisting Nikkie's support against Luit, sometimes by co-opting Luit against Nikkie. This double game began to fall apart toward the end of 1978, when Nikkie and Luit struck up what de Waal calls a "nonintervention treaty." As a result, Nikkie and Luit's special relationship, which appeared during periods of sexual competition, strengthened over the ensuing months.

At the opening of 1980, the situation of the three males seemed stable. Nikkie was the dominant individual, with Yeroen's support, and Luit was excluded, even though he was more powerful than either of the other two. As the year progressed, however, Nikkie seemed lax in keeping his end of the bargain with Yeroen. For instance, he didn't support Yeroen's approach to estrous females, and didn't prevent Luit from approaching females. Two days later, during the night of July 6, a fight broke out in which Nikkie and Yeroen sustained injuries: fingers and toes cut or missing, as is typical of chimp encounters.

"Although no winner or loser could be determined on the basis of mere injury count, Nikkie clearly behaved like the loser," recounts de Waal. "Even though it didn't seem as if Luit had been very much involved in the physical battle, he emerged as the new dominant male." What had happened, believes de Waal, is that because Nikkie had failed to keep his bargain with Yeroen, Yeroen called an end to the alliance in a fairly violent way. When the alliance collapsed, Luit stepped into the power void, once again the dominant male.

During the subsequent weeks the troop was tense because of the fragile order that prevailed after the fight. Luit, Nikkie, and

Yeroen constantly seemed to be testing possible new alliances. But "for Yeroen, restoration of the coalition with Nikkie seemed to have priority over any other option," says de Waal. "Yeroen would scream in apparent frustration and follow Luit and Nikkie around whenever they walked together. And Yeroen himself often tried to sit and groom with Nikkie." It was a patently unstable situation.

Then came the fatal attack. Yeroen and Nikkie appear to have decided that their best interests would be served by re-establishing their alliance. De Waal guesses that Yeroen initiated the attack, leading Nikkie with him. Significantly, the castration that Luit suffered as part of his injuries is not uncommon in status fights of this kind.

The pursuit of power in the Arnhem colony had for the most part been orchestrated with political manipulation and guile, but it ended with the ultimate imposition of the power of an alliance on a rival: death. Quite apart from its tragic elements, I find the story of Yeroen, Nikkie, and Luit arresting. I've been arguing that the world of higher primates—of monkeys, apes, and humans—is quintessentially a game of social chess, a keen intellectual challenge. The challenge is keener yet than the ancient board game itself, because the pieces not only unpredictably change identity—knights becoming bishops, pawns becoming castles, and so on—they occasionally switch colors to become the enemy. To prevail in this game, the player must constantly be alert, ever seeking a winning edge, ever avoiding disadvantage. Chess is an apt metaphor, because it captures precisely the dynamic complexity higher primates face in operating within their worlds. Apt, but somewhat abstract, lacking in emotion.

I wasn't at the Arnhem zoo during the struggle for power that led to Luit's death, but the raw emotion of it all comes through to me just from the description of the events. When de Waal talks about the struggle, he becomes rapt with the personal energy of the participants and clearly is not emotionally indifferent to their ordeal. He admits that when he labels Yeroen a "murderer," he is imputing human motives. I can see why. On a different level we can empathize with the fracas between the vervet monkeys Tycho and Newton over a piece of fruit; cheer on Charing Cross, Newton's sister, as she comes to lend a hand; and understand why Wormwood Scrubs, another of Newton's sisters, takes the opportunity to harass Holborn, Tycho's sister. It makes sense logically and emotionally. And who can avoid a grin of recognition at the shy advances made so patiently by Alex, the young male baboon, to Thalia? The game of social chess in higher primate life is played by individuals pursuing their roles as best they can, expressing a range of emotions that we can identify as human.

What each individual seeks, of course, is reproductive success: producing as many healthy, socially adept offspring as possible. In birds of paradise, the greatest reproductive success (in males) goes to those with the most elaborate plumage and winning display. In red deer, the greatest reproductive success (again, in males) goes to those with the biggest, strongest bodies with which to overthrow rivals, sometimes literally. In higher primates, the greatest reproductive success (in both males and females) is shaped much more by social skills than by physical displays, either of strength or appearance. The complex interactions of the primate social nexus serve as an exquisite sorting system, in which the individuals with an edge in making alliances and monitoring the alliances of others may score significantly higher in reproductive success.

I am describing a world shaped by Darwinian natural selection, so none of the players in it have maximum reproductive success as an immediate, conscious aim in life. Natural selection has sharpened social skills, which lead to reproductive success. Those skills are structured on a keen, analytical intelligence. In other words, natural selection has honed intelligence in primates in the same way and in the same evolutionary context as it influences attributes of strength and physical appearance in other groups of animals.

We began our exploration of the roots of human consciousness by asking why higher primates are more intelligent than they need to be for the daily round of practical affairs. The answer, I suggest, is the intense intellectual demands of primate social interactions, with the constant need to understand and outwit others in the drive for reproductive success. Our brain is extraordinarily large in part because of the exigencies of social interaction, which evolved to levels far greater than those of other higher primates.

17

Consciousness:
Mirror on the Mind

T h e   c o m p u t e r program Deep Thought is widely
regarded as a magnificent achievement of human technological
ingenuity. The result of half a century of unremitting effort, and
the application of the sharpest minds in the computer business,
Deep Thought has scaled the heights to the grand master level
in chess. True, the world champion Gary Kasparov beat Deep
Thought convincingly—two games to nil—in a minitournament
in October 1989. But the program's creators are confident that,
with further refinement, Deep Thought will soon lay claim to
the world's number one spot, certainly by the end of the century.

If a computer can successfully tackle the myriad maneuverings and deep strategies of what may be the world's toughest
intellectual game, what of the game of social chess? Deep
Thought has achieved its grand master status by brute force—or,
rather, speed. Aided by a special chip developed by the system's
creator, Feng-hsiung Hsu, Deep Thought can run through
700,000 possible moves every second. In five minutes it can ex-

amine more than two hundred million possible moves, which, incidentally, are only a tiny fraction of all possible chess moves. Eventually it picks the best one, working about half a dozen moves ahead. Hsu expects Deep Thought to prevail over the world champion when the new chip he is working on runs computation at ten times this speed, looking at seven million moves every second. In the end, brute computing force will triumph over the human brain—on the chess board, at least.

But neither Kasparov's brain nor Yeroen's brain works like Deep Thought's chip. No brain does. That kind of computation would consume too much space and, more significantly, too much time in the slower-acting nerve tissue that makes up the soggy gray mass in our head. Exactly how the human brain or the ape brain or the monkey brain works is, of course, largely a mystery. But it is clear that the human brain employs a lot of clever tricks to reach reasonably good solutions to complex problems without examining every possibility. One of those tricks, developed especially for the challenge of dealing with social interactions, is, I believe, consciousness.

The best way to understand and, more important, to predict the behavior of others under certain circumstances is to know what you would do under the same circumstances. Almost three and a half centuries ago the philosopher Thomas Hobbes made the following prescient statement: "Given the similitude of the thoughts and passions of one man to the thoughts and passions of another, whosoever looketh into himself and considereth what he doth when he does think, opine, reason, hope, fear, &c., and upon what grounds, he shall thereby read and know what are the thoughts and passions of all other men upon the like occasions." It is the use of intuition based on experience.

Inevitably, inexorably, the Inner Eye, as Nick Humphrey calls this mental model, must also generate a sense of self, the phenomenon we know as consciousness: the Inner "I". "In evolutionary terms it must have been a major breakthrough," observes

Nick. "Imagine the biological benefits to the first of our ancestors who developed the ability to make realistic guesses about the inner life of his rivals; to be able to picture what another was thinking about and planning to do next; to be able to read the minds of others by reading his own."

If the mental model produced by the Inner Eye bestows an advantage on individuals in the complex of social interactions, the ultimate goal of which is reproductive success, then it will be favored by evolution. Once established, there is no going back, for individuals less well endowed would be at a disadvantage. Similarly, those with a slight edge would be further favored. "An evolutionary ratchet would be set up," says Nick, "acting like a self-winding watch to increase the general intellectual standing of the species. In principle the process might be expected to continue until either the physiological mainspring is fully wound or intelligence itself becomes a burden."

As humans, we experience the ultimate expression of this dimension of intelligence: the skills of foresight and manipulation, the facility of imagination, the sense of self. We also extend it to raw feelings, of course, to sympathy and empathy, to attribution and affect. This dimension of feeling is what makes consciousness so keenly subjective an experience. An observer may wince in virtual pain on seeing—or even hearing of—an injury to another. A real sense of grief can swell in the emotions of someone who hears of, for instance, a parent losing a child. Empathy with the emotions of others through the experience of one's own emotions is very much part of human consciousness. It also drives the widespread tendency to anthropomorphize, to attribute human feelings to nonhuman animals. The dog "misses" its absent master. The monkey is "jealous" of its rival. The cat is a "selfish" animal. Endowed with this deep sense of self, and living lives awash with emotions, we find it virtually impossible to imagine other lives—any kind of life—without feelings similar to ours.

297

Powerful though our subjective experience of consciousness is, paradoxically it is extremely difficult to prove that it exists at all. As individuals, how can we *know* others feel as we do? How do I know for sure that my neighbor is conscious in the way I know I am? For philosophers and psychologists, it is a tough challenge, although conversation and a resulting empathy may move some distance toward resolving it. But what of nonhuman primates? How can we test whether they too experience a degree of consciousness?

Two decades ago the psychologist Gordon Gallup, now at the State University of New York at Albany, devised a simple, if controversial, test of the sense of self: the mirror test. As many pet owners know, a mirror may be a novelty to a cat or a dog for a while, but once the animal comes to realize that the reflection is at best a boring playmate, it gives up mirror watching pretty quickly. The goal of Gallup's mirror test is to determine whether an animal is able to recognize the reflection as "self" instead of just another individual.

The test is simplicity itself. It involves first familiarizing the animal with the mirror, then marking the animal's head with a red spot. If the animal touches the spot after looking at its reflection anew, then, argues Gallup, the animal does indeed recognize the image as its own. "The first time we tried it with chimps, it worked," recalls Gallup. "These data would seem to qualify as the first experimental demonstration of self-concept in a subhuman form," he wrote in *Science* in January 1970. In the same paper he reported that neither the stump-tailed macaque nor the rhesus monkey "passed" the mirror test.

Since that time many higher primates have been given the test, and so far only two have shown positive results: the chimpanzee, as in the original study, and the orangutan. The gorilla, the third of the great apes, apparently fails, a result that many observers find puzzling. Moreover, some observers claim to have seen self-directed behavior by gorillas in front of mirrors, which they take to indicate the presence of a sense of self in these animals.

Suppose, for a moment, that gorillas do have a sense of self, which, for some reason, the mirror test fails to elicit. In this case there would be a clean line drawn between the great apes and the rest of the higher primates: above the line, a sense of self; below the line, nothing. Such a rigid demarcation has unsettled primatologists for a long time, especially because they see complex social behavior in their nonape subjects, particularly monkeys. What other criterion can be tried? One has emerged recently: deception.

Half a dozen years ago when Richard Byrne and Andrew Whiten were studying chacma baboons in East Africa, they observed the following incident. "Paul, a young baboon, approached and watched an adult female, Mel, who was digging in hard dry earth for a large rhizome to eat. Paul looked around: in the undulating grassland habitat no baboons were in sight, although they could not be far off. Then he screamed loudly, which baboons do not usually do unless threatened. Within seconds Paul's mother, who was dominant to Mel, rushed to the scene and chased her, both going right out of sight. Paul walked forward and began to eat the rhizome."

The two psychologists, intrigued, thought they might have witnessed an instance of true deception. Paul "knew" that if he screamed, his mother would come; she would "assume" that Mel had attacked Paul and would chase her off. He "knew" that he would be left in peace to eat the rhizome Mel had so laboriously unearthed. "There were other interpretations, of course," say Byrne and Whiten. "It may have been sheer coincidence, with the attack by Paul's mother unrelated to his scream. Or Paul may have been genuinely threatened by an adult female not seen by us." But the idea of deception beckoned, particularly as the two psychologists witnessed other occasions where baboons were apparently being "economical with the truth."

On their return from the field Byrne and Whiten did a quick literature search and turned up several reports of "tactical deception," as they call it. Most often, the putative deceivers were chimpanzees. "With their much greater reputation for intelligence, using deception to gain their own ends seemed more in keeping with chimpanzees than it did in a mere monkey," Byrne and Whiten report. "Yet when we excitedly described our precocious animals to other primatologists, for the most part they showed little surprise. Instead, they replied with anecdotes of deception from their own study of animals."

Anecdotes are not generally the stuff of science, and in this case the stories were often open to interpretation. For deception to work in a social setting, it must be close enough to the edge of truth to make detection difficult. It also will not be common, because one can't "cry wolf" often without being caught out. Nevertheless, when Byrne and Whiten canvassed their colleagues' observations, they came up with many instances of putative deception, including such tactics as concealment, distraction, the creation of misleading indications of intent, and manipulation of innocent bystanders. Not only apes, but various species of Old World monkeys (mostly baboons) were cited as accomplished deceivers.

The significance of deception goes beyond its being just another social tool. The agent of the deception must have an idea of what response its action will provoke in the target. The agent must be able to put itself in the mind of the target. In other words, in order to practice deception, an individual must have a clearly developed sense of self. During one of the many maneuverings between the male chimps at the Arnhem zoo, Frans de Waal noticed one day that Nikkie was in a tree while Luit was sitting below. There had been confrontations between the two earlier in the day, and "Nikkie seemed ready for a new display," recalls de Waal. "I noticed that Luit was baring his teeth in a fear-grin. Then I saw Luit pull his lips over his teeth, wiping out the fear-grin. He did it several times. In mutual intimidation between males, it makes sense to hide signs of nervousness. That's what Luit seemed to be doing." The significance here is that Luit seemed able to put himself in Nikkie's head, knowing what Nikkie would think if he saw himself, Luit, wearing a fear-grin.

Research on nonhuman primate deception is by no means clear-cut. The message from Byrne and Whiten's survey, however, is that the facility for tactical deception can be seen in chimpanzees and to a much lesser extent in gorillas. None has yet been detected in orangutans or gibbons, which are much more difficult to study in the wild. Baboons are skilled deceivers. There the line begins to be drawn. Not a single instance of tactical deception was reported in bush babies and their cousins, the prosimians, so it looks as if the phenomenon is real, and that it correlates to some extent with brain size and the complexity of social life.

We are surely seeing here the cognitive foundations of consciousness in our primate cousins, including Old World monkeys. I find it interesting that the foundation appears to be neither deep nor wide, though one would have to say that chimpanzees experience a real elevation in consciousness over Old World monkeys.

How much more conscious are humans than chimpanzees? It is difficult if not impossible to determine. Objectively, we can say that the chimpanzee level of consciousness includes a sense of self sufficiently developed to permit intricate political maneuverings. It allows an individual to place itself in the mind of another so that social chess can be played with considerable skill, including intentional deception. Chimpanzees build models of others' behavior based on their own experience, no doubt. They understand what actions may provoke an angry response, what may elicit fear, and what may elicit friendship. But we might be gliding toward the boundaries of chimpanzee consciousness here, because it is not clear to what degree they experience their raw emotions.

Throwing a temper tantrum is not the same as feeling burning anger. Giving or receiving friendly attention, such as grooming or "kissing," is not the same as feeling happy or feeling in love. Submitting or fear-grinning in the face of threat is not the same as feeling fearful. All animals manifest what we describe as basic emotions, but probably few actually subjectively experience them. The animals would have to be conscious in a way that humans are, to generate sympathy for one's fellows. As far as can be discerned, sympathy—the vicarious experience of emotions—is not well developed in chimpanzees and is even less so in lower primates.

The ultimate vicarious experience, of course, is the fear of death, or simply death awareness. In all human societies, the awareness of death has played a large part in the construction of mythology and religion. There seems, however, to be no awareness of death among chimpanzees. Females have been known to carry around the corpse of an infant for a few days after its death, but they seem to be experiencing bewilderment rather than what we call grief. More important, other mature individuals appear to offer no condolence or sympathy to the bereaved mother. The emotional experience seems to go unappreciated by others, and unshared. So far, no observer has seen reliable indications that chimpanzees have any awareness of the inevitability of their own death, the extinction of self.

What can we say about the direct ancestors of hominids, the ancestors common to us and the African apes? The modern chimpanzee is the product of five or so million years of evolution since that time, of course, so we must beware of equating chimpanzee cognitive capacities with those of all African apes, including the ones of five million years ago. Cautiously, however, I suggest that we can say that large-brained apes that live socially complex lives are likely to develop a chimpanzee level of consciousness. The common ancestor of African apes and hominids falls into or close to this category.

With a starting point of a chimpanzee level of consciousness at the threshold of the human lineage, we can begin to think about the trajectory of its development through human history. The challenge is similar to that of inferring language abilities in our ancestors, only more difficult. The signs of consciousness in the archeological record are even less tangible than those of language. Much of what we can say is only informed speculation.

Luckily, the emergence of a key element of consciousness— the awareness of death—sometimes does make its mark in the prehistoric record. Some kind of ritual occurs, a formalized procedure that identifies and bounds the occasion. From ethnography we know that this may vary from extensive care of the corpse over a long period, perhaps involving moving it from one special location to another after a period of a year or even more, to minimal attention to the body, all effort being devoted to spiritual issues. Sometimes the ritual involves burial, a matter for which prehistorians are infinitely grateful.

The earliest evidence of deliberate burial in the archeological record occurs very late in our history. It comes with the Neanderthals, and presumably with other archaic sapiens populations, not much more than 100,000 years ago. If Neanderthals and other archaic sapiens did have an awareness of death, as I believe they did, what does it tell us about their state of mind and about the evolutionary trajectory of consciousness in human history? Do modern humans have a sharper consciousness than these earlier members of the human family? Piling inference on inference, we can say it is likely that with the origin of modern *Homo sapiens*, subjective consciousness was keener than in archaic sapiens, including Neanderthals. The inference is based on language, and it begins to complete the complex pattern of relations we introduced earlier, between intelligence, language, and consciousness.

Many psychologists and linguists now argue that spoken language is the loom on which some of the finer fabrics of consciousness are woven. The two qualities of the human mind are inextricably meshed with each other. If, as I believe, an enhancement in linguistic facility was a crucial component of the evolution of modern humans, then one might expect a concomitant change in the quality of consciousness.

Since death awareness appears relatively late in our mental development, what can we say of consciousness earlier in our history? What of the mind of *Homo erectus* and of *Homo habilis?* And what of the australopithecines?

First of all, I see no cogent reason for arguing that the chimpanzee level of consciousness we imagine for the beginning of hominid history would have increased significantly in the species before *Homo.* And I view the social structure among these species as having been no more intense than what we see among modern chimpanzees. The mental model of the world produced by the chimpanzee level of consciousness in the australopithecine brains would have been adequate. The evolutionary ratchet of consciousness was not yet far into its inexorable climb.

With the advent of *Homo* and the appearance of the hunting-and-gathering way of life, the game of social chess would have become more demanding. There would indeed have been reproductive advantages to the possession of a more acute mental model, one that would have been aided by a sharper consciousness. Natural selection would have worked with this, moving consciousness to higher and higher levels. This gradually unfolding consciousness not only fashioned a new kind of reality in our heads, it also changed us into a new kind of animal.

The two-million-year heritage of a hunting-and-gathering life, rudimentary at first but ultimately superbly refined, left its mark on our minds just as much as it did on our bodies. On top of the technical skills of planning, coordination, and technology, there was, equally important, the social skill of cooperation. A sense of common goals and values, a desire to further the common good, cooperation was more than simply individuals working together. It became a set of rules of conduct, of morals, an understanding of right and wrong in a complex social system. Without cooperation—within bands, among bands, through tribal groups—our technical skills would have been severely blunted. Social rules and standards of behavior emerged. The great British biologist Conrad Waddington put it best: "Through evolution humans have become the ethical animal."

As we are unsure how big a gap exists between a chimpanzee level of consciousness and our own, we cannot be precise about a *Homo habilis* level of consciousness or a *Homo erectus* level of consciousness. We can only speculate that elements of consciousness —the sense of self, the tendency to attribute feelings to others, the facility to know the world better, and the raw emotion of compassion—all were enhanced through time as the evolutionary ratchet advanced.

I suspect that when the Turkana boy died, his parents experienced grief, had some word for death, some expression for sorrow, and perhaps received sympathy from others in the band. But I doubt whether they understood death in the way we do, as a fate awaiting us all. The apparent absence of death awareness in *Homo erectus* must be taken to imply only a limited facility of self-awareness. I therefore doubt that the Turkana boy's parents would have puzzled over the meaninglessness of his early death or wondered about the meaning of life.

One thing we can be sure about, however, is that once consciousness passed the threshold of self-awareness and death awareness, there welled up in the human mind the Big Question: Why? It is not a straight request for an answer; it is a search for meaning in the midst of uncertainty. What is the meaning of my life? What is the meaning of the world I find myself in? How did the universe come to be? Ubiquitously, there has been the sense that the Truth is somehow unknowable, somehow not meant to

be known. Dostoevsky put it this way: "Man needs the unfathomable and the infinite just as much as he does the small
planet which he inhabits."

As a result, mythology and religion have been a part of all
human history, and, even in this age of science, probably will
remain so. No one has thought or written more extensively
about mythology than the late Joseph Campbell. The lesson of
mythology, he said, is as powerful as it is simple. The elements
of mythology, through space and time, confirm "the unity of the
race of man, not only in its biology but also in its spiritual
history." Perhaps the single most important behavioral adaptation of *Homo sapiens* is the passage from generation to generation
of the elements of culture, the folk knowledge of the means of
survival. Part of that cultural passage is the profoundly felt urge
to understand the world. A people's mythology is its means of
coping with that urge, for mythology is a body of explanation,
an embodiment of the Truth.

It is interesting enough that every human society has felt the
need to generate a body of myth, an explanation of how the
society came to be and its place in the world. Even more interesting are the many commonalties between different mythologies. "The comparative study of the mythologies of the world
compels us to view the cultural history of mankind as a unit,"
observed Campbell. "We find that such themes as the fire-theft,
deluge, land of the dead, virgin birth, and resurrected hero have
a worldwide distribution—appearing everywhere in new combinations while remaining, like the elements of a kaleidoscope,
only a few and always the same."

The way people arrived at answers about their world followed much the same path individuals take in coming to understand one another. In all of the mythologies that we know, and
by extrapolation in mythologies long extinct, many of the important elements, such as animals and physical forces, are endowed with humanlike emotions and motives. The mind that

evolved subjective consciousness as a tool with which to understand the complexities of social chess used the same formula to understand the complexities of the rest of the world. It is anthropomorphizing on a cosmic scale.

To the earliest members of *Homo sapiens*, and to societies through much of human history, life was played out in full interaction with other powers in the world. The interaction assumed, if not fully human qualities in these powers, then at least some human qualities. The migratory herd had to be treated with respect; otherwise it would refuse to return next season. Appropriate gifts had to be made to the sun; otherwise it would become angry and not rise. The spring had to be constantly blessed; otherwise it would choose to flow elsewhere.

Explanation, then, was what people sought, not as demonstrated fact but as authorized story, the basis of myth. The definition of myth, according to Alan Dundes, an anthropologist at Berkeley, is "a sacred narrative explaining how the world or humans came to be in their present form." The origin myth is the most fundamental story of all societies, and every society has one. Not only does the origin myth serve the purpose of telling how a particular society came into being; it also explains and therefore justifies the nature of that society. Consider the Yanomamo Indians, whose territory straddles the border between southern Venezuela and northern Brazil. For these people, warfare and violence is a way of life. Napoleon Chagnon, an anthropologist at Northwestern University, who has studied the Yanomamo for many years, refers to them as "the fierce people." The Yanomamos' origin myth encompasses this aspect of their life. Yanomamo elders told Chagnon their origin myth, and he recounts it as follows.

After the flood, there were very few original beings left. Periboriwa (Spirit of the Moon) was one of the few who remained. He had a habit of coming down to earth to eat

307

the soul parts of children. On his first descent, he ate one child, placing his soul between two pieces of cassava bread and eating it. He returned a second time to eat another child, also with cassava bread. Finally, on his third trip, Uhudima and Suhirina, two brothers, became angry and decided to shoot him. Uhudima, the poorer shot of the two, began letting his arrows fly. He shot at Periboriwa many times as he ascended to *hedu*, but missed. People say he was a very poor shot. Then Suhirina took one bamboo-tipped arrow and shot at Periboriwa when he was directly overhead, hitting him in the abdomen. The tip of the arrow barely penetrated Periboriwa's flesh, but the wound bled profusely. Blood spilled to earth in the vicinity of a village called Hoo-teri, near the mountain called Maiyo. The blood changed into men as it hit the earth, causing a large population to be born. All of them were male; the blood of Periboriwa did not change into females. Most of the Yanomamo who are alive today are descended from the blood of Periboriwa. Because they have their origins in blood, they are fierce and are continuously making war on each other.

The story goes on to explain that women originally sprang fully formed from the body of one of the men. But, says Chagnon, the essential point is that "this myth seems to be the 'charter' of Yanomamo society." The fierce people are fierce because of their origins. The same pattern is to be found in all origin myths: they describe both the origin of the people and the nature of their world. Explanation is both descriptive and prescriptive. It provides a framework for life.

The appearance of a devastating flood in the Yanomamos' origin myth is, incidentally, just one of many examples of a flood as an essential agent in a society's birth. Real floods can loom large in the worlds of many people, and often must have threatened their safety. But the ubiquity of flood myths—they can be found in societies on every continent—has convinced anthropologists that their origin is more fundamental, less tangible. "I would ascribe these myths to that basic and clearly universal human longing—manifested less dramatically when a man changes his job or moves to a new house—to get rid of an unsatisfying past and start all over again," speculates the anthropologist Penelope Farmer. "Just so a world, postflood, could be restored to innocence, to another Eden, all bitter experience laid aside, and the history of mankind begin anew."

Animals figure large in many societies' mythologies, not surprisingly, as hunter-gatherers rely heavily on animals as resources. They were anthropomorphized in terms of their "intentions," and often took on special roles in people's interaction with "spirit worlds," sometimes representing sources of power. Frequently, animal images are distorted, becoming part human and part beast, an expression of the ambiguity of life, an elision of human, animal, and spirit worlds.

The ultimate expression of this anthropomorphism, of course, is the creation of gods. "The Old Testament states that God created man in His own image," note Gordon Gallup and Jack Maser. "We would argue that the opposite has occurred. Because of our capacity to use personal experience as a means of understanding the experience of others and because of the well-studied phenomenon of generalization, humans create God(s) in their own image, and not vice-versa."

Maser, a researcher at the National Institutes of Health, in Bethesda, joined Gallup in a recent essay, "Theism as a By-product of Natural Selection," in which they extended Gallup's ideas on human consciousness to reach the product of human attribution. "In another reversal of a familiar idea, we would say that it is awareness of self that should be construed as a high-level abstraction; God then follows as a rather concrete extension of self."

Our discussion of consciousness began with the death of a chimpanzee; touched on a computer that can play grand master
chess, and finishes here with God as the creation of the human
mind. Consciousness, as a quality of mind, makes each of us feel
special as an individual, because the sense of self, by its nature, is
exclusive of others. The same quality has encouraged us—*Homo
sapiens*—to feel special in the world, separate from and somehow
above the rest of nature.

The evolution of human consciousness was the fourth great
biological revolution in the world, a new dimension of biological
experience: the self having become aware of itself. With the
birth of consciousness was also born the urge to know, in both
tangible and intangible realms. One only has to look around at
the material world we inhabit—a world we created—to see the
impact of human consciousness on the world. Great science,
great art, and great compassion—each is the product of consciousness. And great arrogance.

Having been seduced into believing that we are indeed special in the world, we have come to take an anthropocentric view
of the world—and, for many people, of the universe too. An
anti-anthropocentric critic—perhaps a visitor from a civilization
far more advanced than ours—might point out that the quality
of consciousness of which we are so proud is in fact a fragile
entity, a cognitive illusion created by a few neuronal tricks in the
midst of gray matter. I won't stray into this treacherous philosophical territory, but it is worth thinking back to Colin Mc-
Ginn's warning.

He suggests that, although we may not like the idea, the
human mind must have limits to what it can comprehend, and its
own consciousness may be one of these things beyond those
boundaries. There may be other things too. "It is deplorably
anthropocentric to insist that reality be constrained by what the

human mind can conceive," he says. "The limits of our minds are just not the limits of reality." There must be realities beyond our own in the universe, a phenomenon future generations may have to cope with.

Meanwhile, here on Earth, paleoanthropology teaches us that our reality is rooted in our history, linked by other, different realities through an unbroken chain of ancestors to an unconscious past.

18

Windows
on Other Worlds

The smell of the burning oil lamp seized a dim memory trace and whirled my mind back to my childhood, to a visit when I was about five or six years old to Lascaux, the most spectacular of all the painted caves of the Ice Age. Brief utterances hushed in awe as my mother, my father, and the Abbé Breuil, France's most famous archeologist, spoke about the images on the walls in front of us. My parents had been working with prehistoric rock art in eastern Africa and were eager to see those in Europe, and to discuss them with the abbé. It was an intense experience, for everyone, this visit to Lascaux. I knew I was in the presence of something quite remarkable, but wasn't aware of what it was. I don't remember what was said. I don't even remember which paintings they looked at. I just remember the sense of reverence, veneration that quieted their voices. And the smell of the burning oil.

I returned to France in 1980, three decades after that childhood visit, to work on a film series with the BBC. The series, called "The Making of Mankind," included an episode on Ice Age art, so I was to have another opportunity to see some of the most outstanding and most arresting relics of human prehistory. Graham Massey, the BBC producer for the series, had arranged for us to film in several caves, some in the Dordogne and some in the French Pyrenees, a most specularly beautiful region of Europe. In the first of them, La Mouthe, near the town of Les Eyzies in the Dordogne, we were met by Monsieur Lapeyre, the owner of the cave, who had agreed to be our guide. He led us into the mouth of the cave, the bright sunlight throwing stark shadows and soft reflections on the rock surface. Soon the light thinned and the gloom gathered. That was when Monsieur Lapeyre lit his old-fashioned oil lamp.

We made our way along a narrow gallery, the lamplight flickering on the wall and ceiling of the ancient cavern. Suddenly we stopped, and Lapeyre pointed. *"Voilà!"* Before us was the beautifully engraved figure of a bison, with great curving horns. it was almost as if my parents and the Abbé Breuil were standing by my side, so powerful were the emotions fueled by those memories. And the sense of awe I felt at standing before this arresting image was just as strong as it had been three decades earlier. Here, seventeen thousand years ago, someone had transferred something from his or her mind onto this wall. The event must have been extraordinary in some way, one feels sure of that, so imbued with meaning does it seem. If La Mouthe had so profound an impact on me, what, I wondered, would happen when I saw Lascaux again? It would not be long before I found out.

These prehistoric images speak to us more evocatively than any other element of the archeological record: colorful, vibrant paintings of horses, of bison, of a panoply of animals and humans that often seem alive and in motion. And yet there is a dimension of unreality about them, for these are not scenes of the Ice Age. The images seem plucked from life, to become part of the cave wall, often arranged chaotically to our eye, frequently superimposed one upon another, sometimes apparently incomplete. Some of them are like monsters, part animal, part human. There is an enigma in these images, a profound challenge to our understanding of our past.

Ever since prehistoric art was discovered in the late nineteenth century, it has held archeologists in its thrall: the quest for what the images mean has been constant. By now more than two hundred painted caves are known in Europe, mostly in France and Spain. In addition to the painted images in these caves, less obvious, but usually far more numerous, are engraved and carved images. Because they are less spectacular than the paintings, the engravings often receive less attention. But they were obviously a major component of the people's culture here, between thirty-five thousand and ten thousand years ago, the era in Europe known as the Upper Paleolithic. So too were the delicately engraved and carved objects, such as spear throwers and hide scrapers, known collectively as portable art. Upper Paleolithic people also manufactured beads and pendants, which were used to decorate clothing and their bodies.

To our Western eyes, the painted images are the most prominent component of a corpus of artistic expression. This Western bias, a particularly Eurocentric bias, has been pervasive and deep. It has fostered unconscious modern European perspectives on the meaning of prehistoric art in Europe, and it has resulted in a lack of attention to, and concern about, prehistoric art of equal and sometimes greater antiquity in eastern and southern Africa.

My parents spent many years finding and copying paintings on scores of rock shelters in Tanzania. I'm proud to say that I recently helped Mary publish some of the best examples of these paintings, a poignant record of a fast-disappearing part of our history in Africa. In southern Africa, much of the equally fragile rock art of the San people has now been recorded and is being carefully studied. Because the art of Africa was on rock shelters, not in deep caves, as in Europe, the ravages of time have eroded most of a rich artistic expression. What we see now is the merest glimpse of what was on the minds of these people.

Before we explore some of the history of this most remarkable aspect of the human record, a word of caution is necessary. Each society weaves its own culture, a complex fabric of many elements, each element giving special meaning to the others. It is often difficult for someone from outside a particular culture to understand the fabric as an entity. Differences in language, in values, and in mythology create barriers to understanding. Pluck a single thread from that fabric, and the foreigner is even less likely to comprehend its significance. The painted, engraved, and carved images of prehistory are threads from past cultures, and we are the foreigners trying to interpret their meaning. Perhaps more than anything else, art can be fully understood only in the context of the culture that produced it.

The first major discovery of prehistoric art was the Spanish cave of Altamira, which, like Lascaux, is one of the most spectacular examples of Upper Paleolithic art yet known. As has often happened in prehistory, Altamira's claim to be a part of our past was initially met with skepticism.

The old farm of Altamira, meaning "high lookout," is situated on gently sloping but high meadowland some three miles from the Cantabrian coast, in northern Spain. To the south, the Cordillera Cantábrica dominates the skyline, and the frequently snow-clad Picos de Europa rise to heights of almost ten thousand feet in the west. It is a dramatic setting. Beneath the farm, caverns and narrow corridors snake through the soft limestone. That the area was riddled with caves was common knowledge, but until 1868 the cave of Altamira was unknown to the owner of the land, Don Marcellion de Sautuola. In that year a hunter came across the cave entrance as he attempted to rescue his dog, which had fallen among some rocks while chasing a fox.

De Sautuola was something of an amateur archeologist, so when he heard of the discovery he explored the new cave briefly; he found very little to interest him. Ten years later, on a visit to Paris, he talked to the famous French prehistorian Edouard Piette about life in the Ice Age. Piette advised de Sautuola how to explore cave sites more efficiently. It was therefore with more enthusiasm and more expertise that de Sautuola returned to the Altamira cave in 1878. He eventually found that it meandered through a cavernous system of more than three hundred yards. But his archeological haul was still meager; it comprised just a few stone tools.

The treasure of Altamira might have remained hidden forever had it not been for de Sautuola's small daughter, María. One day in 1879 María accompanied her father to the cave and wandered into a low chamber that de Sautuola had explored previously. Whereas her father had had to crawl through the chamber, María could stand up in it. She looked at the ceiling and, in the flickering light of an oil lamp, saw images of two dozen bison grouped in a circle, with two horses, a wolf, three boars, and three female deer around the periphery, images in red, yellow, and black, as fresh as if they had just been painted. María's father was astonished to see what he had missed and his daughter had found. He knew it must be a great discovery.

When de Sautuola visited Paris in 1878, he had seen at the Grand International Exhibition a display of engraved stones collected from a number of French caves, which scholars had accepted as being prehistoric. He saw in the images in Altamira echoes of those engravings. His excitement and delight can be imagined, as can his shock and disappointment when scholars dismissed the paintings as nothing more than the work of a modern artist. They looked too good, too realistic, too artistic to be the work of primitive minds. One French scholar even suggested that an artist who had stayed with de Sautuola for some time had done the paintings. Appalled by the scholars' reactions, containing more than a hint of fraud, de Sautuola closed the cave. He died in 1888.

De Sautuola did have some supporters, however, most notably Piette, who had encouraged his searches. A year before de Sautuola died, Piette wrote to Émile Cartailhac, the leader of the opposition to Altamira's authenticity, urging him to reconsider his position. The appeal failed, and Altamira had to wait another fifteen years for recognition. The acceptance of Altamira was brought about by a steady accumulation of similar finds, albeit of lesser impact. First, La Mouthe, a cave with painted and engraved bison and a fine example of a stone lamp, was discovered in the Dordogne in 1895, its dating to the Ice Age incontrovertible. (This was the first cave I visited in 1980.) More discoveries followed, in Font-de-Gaume and Les Combarelles, both in the Dordogne, for example. The weight of evidence built until it proved persuasive. Cartailhac admitted his mistake in a famous essay, "Mea Culpa d'un Sceptique," published in 1902.

Once the archeological establishment accepted that artistic expression was to be found in the Upper Paleolithic, intellectual energies turned to trying to understand what the images meant. One early, popular notion was that they were simply art for art's sake. John Halverson, an anthropologist at the University of California, Santa Cruz, has recently revived this hypothesis, suggesting that the images are products of the "primal mind," and that we are witnessing "human consciousness in the process of growth." One of his arguments, paradoxically, is that the images are clear representations of animals of the time. "Paleolithic representation is naturalistic because it is unmediated by cognitive reflection," he says. "At this early stage of mental development, percept and concept may have been undifferentiated." The images, he claims, are simple, depict no scenes, and show "nothing that can be attributed with any solid assurance to religious motivations."

On this latter point, given the caveat about viewing the art

317

through Western eyes, one might justifiably ask how one can be certain whether a particular set of images had religious motivations. As Upper Paleolithic religion and mythology probably were different from anything we know now, we are unlikely to recognize as significant something that to Paleolithic people might, for example, encapsulate an entire origin myth. What is more pertinent to the argument is that the images are not as simple as Halverson implies.

It is true that, with rare exceptions, the engravings and paintings do not represent scenes from Upper Paleolithic life. Plants are only rarely depicted, so there is nothing like a true landscape painting. Nor are there many animal scenes that are truly naturalistic. But there is a pattern. Count the images of horse and bison, and one accounts for a large proportion of all the images. Add oxen, and some 60 percent of all the images are accounted for. Mammoth, deer, ibex, boar, rhinoceros, and goat also appear, but in small proportions. So too do fish and birds. Carnivores like lions, hyenas, foxes, and wolves are rare. There is no doubt that the images as they appeared on the walls did not accurately represent the animals in nature: some, by their strength in numbers, seem to be accorded more importance.

For these and other reasons, the early idea of art for art's sake was soon abandoned, to be replaced by the hypothesis of sympathetic magic, or hunting magic. At the turn of the century, anthropologists working in Australia were beginning to realize that the Aboriginal paintings were part of magical and totemic rituals. The same could be true of Upper Paleolithic art, argued Salomon Reinach in 1903. The Australians and the Upper Paleolithic people were both hunter-gatherer societies, he noted. Both societies produced paintings in which a few species were overrepresented in relation to the natural environment. Reinach reasoned that Upper Paleolithic people made paintings to ensure the increase of totemic and prey animals, just as the Australians were known to do.

Reinach was not the first to invoke hunting magic as the motive underlying Upper Paleolithic art, nor was he the most famous name to be associated with it. The Abbé Breuil was convinced by Reinach's ideas, and he developed and promoted them vigorously during his long and dominating career. Cartailhac had invited Breuil to Altamira the year after he published his "Mea Culpa," and Breuil, twenty-six at the time and already an expert on the end of the Ice Age, became entranced with the Age of Art that preceded it. For almost sixty years, Breuil recorded, mapped, copied, and counted images in the caves throughout Europe. He also developed a chronology for the evolution of art during the Upper Paleolithic period. During all this time Breuil continued to interpret the art as hunting magic. So did the majority of the archeological establishment.

The Abbé Breuil died in 1961, and with him died the all-encompassing hunting-magic hypothesis. By this time another French archeologist, André Leroi-Gourhan, had been developing his own interpretation, one based on the emerging ideas on structuralism. Where Breuil had seen chaos—or, at least, randomness—in the wall art, Leroi-Gourhan sought and found order. "Indeed, consistency is one of the first facts that strikes the student of Paleolithic art," he said. "In painting, engraving, and sculpture, on rock walls or in ivory, reindeer antler, bone, and stone, and in the most diverse styles, Paleolithic artists repeatedly depict the same inventory of animals in comparable attitudes. Once this unity is recognized, it remains only for the student to seek ways of arranging the art's temporal and spatial subdivisions in a systematic manner." This order, he suggested, embodied ideas about Upper Paleolithic society.

One problem with the hunting-magic hypothesis was that the images depicted very often did not reflect the diet indicated by fossil remains. Reindeer were a large component of the diet, but paintings of reindeer were not common. The reverse was true for horses. As Claude Lévi-Strauss once observed of art among San and Australian Aborigines, certain animals were depicted frequently not because they were "good to eat" but because they were "good to think." The question is what the thinking was about. Leroi-Gourhan's answer was that it had to do with the structuring of society, the division between maleness and femaleness.

The horse image represented maleness in Leroi-Gourhan's scheme, and the bison, femaleness. The stag and the ibex were also male; the mammoth and the ox were female. Leroi-Gourhan surveyed more than sixty caves and saw order in the distribution of their images. Deer, for instance, often appeared in entranceways but were uncommon in main chambers. Horse, bison, and ox were the predominant creatures of the main chambers. Carnivores mostly occurred deep in the cave system. Although he modified details of his male-female duality system later on, Leroi-Gourhan always saw structure as an important element in the art, something that persisted through space and through time.

Another French archeologist, Annette Laming-Emperaire, also saw structure in the distribution of images, also as a male-female duality. Unfortunately, in some cases what one prehistorian inferred as maleness, the other saw as femaleness. And vice versa. These differences of opinion "dealt a trump card to the critics of this new vision of Paleolithic art," lamented Laming-Emperaire. In the end, though, it wasn't so much problems of this sort that undermined the Leroi-Gourhan hypothesis; rather, the hypothesis was too all-embracing, too monolithic. "Archeologists began to look at the diversity in the art," explains Margaret Conkey of the University of California at Berkeley. "There was a concentration on a diversity of meanings and a concern for the context of the art, not so much what the images meant but rather what made them meaningful."

Leroi-Gourhan died in 1986, just as his ideas were being replaced by the approach based on diversity. The second great era of the study of Paleolithic art was therefore at an end. Since then, no single figure has emerged to dominate the field in the way that Breuil and Leroi-Gourhan did. For instance, Denis Vialou, of l'Institut Paléontologie Humaine in Paris, sees order in the distribution of images within caves, but not the global order that Leroi-Gourhan claimed. "Each cave should be viewed as a separate expression," he says. Meanwhile, Henri Delporte, of the Musée des Antiquités Nationales, near Paris, is concentrating on the differences in pattern of wall art and portable art. Conkey is focusing on clues to the social context of the production of images. And so on. It is a time for change in the study of Paleolithic art, a search for new ways of looking through that window into the Paleolithic mind.

Before we develop some of these recent ideas more fully, we should sketch the overall picture of Upper Paleolithic art. The period covers the time thirty-five thousand to ten thousand years ago, its termination coinciding with the end of the Ice Age. Five cultural eras have been identified in Europe during this time, based principally on changes in technology. They are: the Aurignacian, thirty-five thousand to thirty thousand years ago; the Gravettian, thirty thousand to twenty-two thousand years ago; the Solutrean, twenty-two thousand to eighteen thousand years ago; and the Magdalenian, eighteen thousand to ten thousand years ago. Although these cultural eras are based primarily on technological innovations and characteristics, elements of the art change too.

The phrase "Upper Paleolithic art" may seem to imply a uniformity, both in style and through time. There is none. Although there are certain consistencies in style, such as the importance of horse and bison as painted images throughout the period, there is great diversity too, both in space and time. And the most famous images, pictures from Lascaux and Altamira, for example, and the elegantly carved spear throwers from La Madelaine, all come from the last of the five eras, the Magdalenian. In quantitative terms, some 80 percent of all Paleolithic art comes from the Magdalenian.

So impressive is the art of the Magdalenian that, perhaps inevitably, it has been called the high point of Upper Paleolithic art, as if there had been an artistic school lasting twenty-five thousand years, constantly seeking to better itself. Again, the Magdalenian looks particularly impressive to us because it is closest to a Western concept of "art." Leroi-Gourhan made the link explicit and called the Magdalenian "the origins of Western art." This is clearly not the case, because at the end of the Magdalenian representational painting and engraving all but disappeared, in the so-called Azilian era, to be replaced by schematic images and geometric patterns. Many of the techniques that had been applied in Lascaux, such as perspective and a sense of movement, had to be reinvented in Western art, with the Renaissance.

The first era of the Upper Paleolithic, the Aurignacian, is notable for several reasons, one of which is the absence of painted caves. The production of ivory beads for body ornamentation was important, and so too was the manufacture of small, carved human and animal figures. From the site of Vogelherd in Germany came half a dozen tiny mammoth and horse figures sculpted from ivory. One of them, an exquisite horse figure, is as skillfully produced a piece as can be found throughout the Upper Paleolithic. There are bone fragments and ivory plaques festooned with regularly spaced incisions, perhaps simple decoration, perhaps, as Alexander Marshack has suggested, systems of notation. One of the most evocative pieces from this era, from the Abri Blanchard in southwestern France, is a flute. Music must have been an integral part of Upper Paleolithic life too.

During the Gravettian, the second era of the Upper Paleolithic, artistic expression included more media. For instance, clay figurines—some animal, some human—were made at a site in Czechoslovakia. Negative handprints are found in some caves; they were made perhaps by holding the hand up to the cave wall and blowing paint around the edges. In the site of Gargas, in the French Pyrenees, more than two hundred prints have been counted, almost all of them missing one or more parts of fingers. The most characteristic innovation of this era, however, are the female figures, often lacking facial features and lower legs. Made from clay, ivory, and calcite, and found throughout much of Europe, they have typically been called Venuses—that Western projection again—and have been assumed to represent a continentwide female fertility cult. In fact, as scholars have recently come to realize, there is a great deal of diversity in the form of these figures throughout Europe, and few would now argue for the fertility cult idea.

It is only with the third era, the Solutrean, that painting in deep caves began to appear in any importance, and then only secondarily so. Much more significant in this era was the development of the carving of large, impressive bas-reliefs, often situated at living sites. One of the most exceptional was at the site of Roc de Sers, in the Charente region of France. Large figures of horse, bison, reindeer, mountain goat, and one human line the back of the rock shelter, some of them standing out fifteen centimeters or so in relief.

And then there was the Magdalenian, the era of deep-cave painting, of Altamira and Lascaux; of exquisitely carved and engraved ivory objects, some utilitarian, such as spear throwers, some not obviously so, such as "batons"; and of a burst of depiction of human faces, for instance at the cave of La Marche, where a veritable picture gallery of more than a hundred individuals was engraved in limestone blocks.

What can the art tell us about the Upper Paleolithic as a whole? First, the persistence and widespread production of especially emphasized images—particularly the horse and the bison

323

—in the painted caves must be significant. It suggests to me a community of bands in constant contact with each other, linked by trade and shared traditions. To Randall White, the evidence of trade is cogent. "Many people think of Upper Paleolithic societies as small, self-contained units," says White, of New York University. "But there's a lot of evidence for the exchange of items across large distances. In some sites in the Ukraine, for instance, you find seashells of a sort that come only from the Mediterranean. You find amber in sites in southern Europe, and it must have come from northern Europe, near the Baltic Sea." There are many such examples, says White, and they are best interpreted as the exchange of items between different groups.

"In the modern world, we tend to think of exchange, or trade, as a purely economic transaction," he explains. "But in most small-scale societies, it operates as a vehicle of social obligation . . . Obligations are social bonds capable of tying together different social groups." This kind of alliance building between social groups, so important among hunter-gatherer bands, is the most sophisticated expression of the social chess we saw in nonhuman primates, where alliances were mostly between individuals.

Alliances between modern hunter-gatherer bands are maintained and strengthened during occasional aggregations of the bands; sometimes many bands come together at the same time. The bands often have different reasons for aggregating at certain times of the year. For instance, bands of the !Kung San of southern Africa aggregate during the rainy season, when new waterholes form. Their neighbors, the G/wi San, aggregate in the dry season, when only a few waterholes are available. Different rationales, but in each case the aggregation is an opportunity for renewing friendships, strengthening political alliances, and arranging marriages. The pattern, common among recent hunter-gatherer societies, may have been present during the Upper Paleolithic.

Margaret Conkey has suggested that Altamira may have been an aggregation site, a place on which neighboring bands converged during the fall, perhaps, when red deer and limpets were plentiful. But the real benefit of the aggregation would have been social and political, not economic. The arrangement of groups of animals on the painted ceiling of Altamira may even reflect the different outlying bands, Conkey speculates. One clue here is that the tools found at Altamira embrace the range of tools found at different sites in the region. Unfortunately, there is no such archeological evidence to suggest that Lascaux was a major aggregation site. But it has to be said that the survey for open-air sites in the region has not been great.

Jacques Marsal was a performer at Lascaux, the Lascaux of the modern age, that is. Marsal, one of the four small boys who discovered the cave by accident in 1940, served for many years as a guide there. He delighted in leading his small groups of visitors into the dark cave, a thin beam from a flashlight and a handrail the only indication of the route to follow. Marsal made the most of the dramatic moment, building anticipation. The trick worked with me. How could it fail, being seconds away from seeing the greatest treasure of the Ice Age?

Marsal would wait until there was absolute silence and then throw a switch, when light would flood the huge chamber. There is truly no easy way to explain what one sees or feels there, in the midst of visual bombardment by chaotic activity. The images have such an urgent presence and energy that the sound of their hoofs and the smell of their hide press through the silence of the cave. On the left-hand wall, a cavalcade of prehistoric beasts stampedes toward the deeper recesses of the chamber. Four gigantic white bulls in black outline dominate the long cavern where it widens to form a rotunda; this is the Hall of Bulls. A menagerie of smaller creatures jostles among the legs of the great beasts. Trotting horses, tense stags, and frisky young ponies stand out from the walls and ceiling in black, red, and

yellow, sometimes bold drawings, sometimes just tantalizing suggestions. Several images overlap others, some are huge, some
diminutive. A fine purple-red horse with a rich, flowing black
mane hangs near two great bulls facing each other in head-on
challenge.

Standing in the Hall of Bulls, surrounded by this wild scene,
which exudes so much vitality and power, one is overwhelmed
by the sense of another age. During my career I have been privileged to handle many of the great relics of human prehistory,
some of which I've had the good fortune to excavate myself. The
sensation of connectedness that I've always experienced with
them has been profound; they are precious links with our past.
Here, in Lascaux, the emotional charge I felt challenged all that.
Next to the discovery of the Turkana boy, the visit to Lascaux
ranks as one of the great moments of my life.

From the Hall of Bulls are two exits, each extraordinary. The
first leads to the Axial Gallery, a narrow passageway richly decorated on all sides: a large red cow with a black head, a stag in full
roar, several delightful yellow horses, and finally a large horse
galloping toward the end of the gallery, where the calcite-
encrusted ceiling swoops down and forms a pillar. There, winding itself around the pillar, lying on its back, its legs flailing the
air, mouth open as if whinnying, is a horse. In order to see the
image, I had to contort myself into this cramped space. Imagine
the challenge of painting it in the first place. Imagine the motiva-
tion to paint it there and in that position.

Imagine. The yellow flame of the oil lamp flickers in the
cold, still air of this corner of the great cave, the young man
holding the lamp trying in vain to throw good light onto the
surface of the pillar. Echoing down the narrow gallery, the sound
of rhythmically stamping feet and urgent, repetitive song. The
atmosphere tense, everyone by now exhausted yet taut with expectation. They know the event is near an end. For an hour, the
undulating intensity of their dancing, their singing, their emo-

tion, had swept them onward, seeing with unwatching eyes as the shaman surged deeper and deeper into a trance, entering a world forbidden to all others. Trembling, groaning as if in pain, with eyes turned inward to that other world, the shaman, clad in horsehide, has left their circle, gone down the narrow gallery, drawing power from the images already on the walls by turning to them, touching them, being them; now to serve once more as the medium for the equine potency.

Crouched by the pillar, with paints provided by the youth, the shaman works at the image with a fervor whose source was another world, the world of the equine potency. The spirit knows what image it needed; the shaman is merely the channel, the contorted rock face not just an allegory of despair, but now the experience of despair. No longer is the end of this gallery an interesting rock formation; it has been transformed into the Place of Despair. The equine potency knows why. The people have an inkling, and talk in whispers about it. But even if she could remember the trance experience, remember being the whinnying horse facing disaster, cipher for the fears of the coming year, she wouldn't talk about it. Shamans don't.

My fantasy, of course. But in these painted caves, in Lascaux especially, the images and the often inexplicable situation drive the imagination. There is a tangible power to these places that speaks of their importance in the lives of our ancestors.

The second exit from the Hall of Bulls leads to a different Lascaux, just as enigmatic. A long, winding cavern stretches eighty yards, in places difficult to pass, almost all of it extraordinarily decorated. A long section at the beginning is covered with a profusion of engravings, some tiny, some enormous, often incorporating features of the rock surface in the visual effect. One, for instance, has a tiny lump for its eye, another a bulge as a three-dimensional effect to a stag's abdomen.

The furthest point of the cavern, and least accessible, is the Chamber of Felines, a place visitors rarely reach. Its inaccessibility speaks of a very special respect for the animals depicted there, lions. It must be significant in this context that when Upper Paleolithic people used animal teeth as items for pendants, they almost always used the teeth of carnivores—lions, hyenas, wolves.

No description of Lascaux is complete without the story of the Shaft. Just off the Apse, halfway along the passageway toward the Chamber of Felines, is a twenty-foot-deep hole, wide enough for one person to be in comfortably, but little more. A metal ladder is the way down, and a flashlight beam illuminates the scene. Among the sparkling yellow and white calcite crystals on the wall, a great black bison is poised for attack, its forelegs taut as a spring, its tail lashing. The animal appears to be desperately wounded, with what looks like a barbed spear across its body. Its entrails spill to the ground. A man has fallen in front of the bison, not a figure crafted with the fidelity of the other images in Lascaux, but a stickman with no life, wearing what might be a bird mask. Nearby, what looks like a bird on the end of a staff, perhaps a spear thrower, and a rhinoceros. All painted in black, and about as enigmatic as anything in Lascaux.

The most obvious interpretation of the scene in the Shaft is that it is connected with hunting magic, perhaps the re-enactment of a hunting accident. But the most obvious explanation may not be the correct one, for three pairs of dots separate the rhinoceros from the rest of the scene. Simple in themselves, and perhaps without import, the dots are just one example of an element in Lascaux art, and in all cave art, that I have not yet mentioned. This is the profusion of nonrepresentational, geometric patterns. In addition to dots, there are grids and chevrons, curves and zigzags, and more. Many kinds of patterns are to be found, sometimes superimposed on animal images, sometimes separate from them. The coincidence of these geometric motifs with representational images is one of the most puzzling aspects of Upper Paleolithic art.

For the Abbé Breuil, these geometric patterns, or signs, as they are called, were part of hunting paraphernalia: traps, snares, even weapons. Leroi-Gourhan included them in his structural duality. Dots and strokes were male signs, he said; ovals, triangles, and quadrangles were females signs. Just recently a South African archeologist, David Lewis-Williams, has suggested that neither interpretation is correct. They are, he says, images plucked from a mind in the state of hallucination, a sure sign of shamanistic art. His argument is based on a study of San art, in southern Africa, and a neuropsychological model that may be basic to much human image making in hunter-gatherer societies, including those of the Upper Paleolithic.

When Lewis-Williams began studying San art four decades ago, it was generally interpreted as representing simple, schematic images of everyday San life. Recently he realized that the images were not realistic in that sense but instead were shamanistic art, which has a different kind of reality, the reality of another world. The key insight here has to do with the trance-induced hallucination that shamans experience during their rituals.

Having made the link between the art, the shamans, and hallucination, Lewis-Williams resorted to the neuropsychological literature, seeking clues to that connection. "There were reports of visual hallucination, very precise descriptions," he says. "The research shows that in early stages you see geometric forms, such as grids, zigzags, dots, spirals, and curves." These images, six different kinds in all, are shimmering, incandescent, mercurial—and powerful. Called entoptic images—which means "within vision"—these phenomena are products of the basic neural architecture of the human brain. "Because they derive from the human nervous system, all people who enter certain altered states of consciousness, no matter what their cultural background, are liable to perceive them," says Lewis-Williams.

In a deeper state of hallucination, stage two, people try to make sense of these images. The results are dependent on an individual's culture and present concern. A series of curves may be depicted as hills if the subject is thinking about the country-side, for example, or waves on the sea if he has thoughts of sailing. San shamans frequently manipulate series of curves into images of honeycombs, since bees are a potent symbol of super-natural power that these people harness when entering a trance.

People who pass from stage two hallucination to stage three often experience a sensation of a vortex or rotating tunnel around them, and soon have hallucinations filled with iconic images, not just signs. "While Western subjects hallucinate air-planes, motorcars, dogs, and other animals familiar to them," says Lewis-Williams, describing laboratory experiments, "San shamans hallucinate antelope, felines, and circumstances, though bizarre and terrifying, derived ultimately from San life." In this final stage, subjects come to "inhabit rather than merely witness a truly bizarre hallucinatory world." It is here that "monsters" ap-pear, part human, part beast, known as therianthropes.

Having established this three-stage neuropsychological model, Lewis-Williams, together with his colleague Thomas Dowson, turned again to San art to see how it might fit. "The first thing we found was that all six entoptic signs are in San art," recalls Lewis-Williams. "This encouraged us to believe that the model was valid, because we knew that shamanism was impor-tant in San life." Indeed, there was good ethnographic evidence that San art was shamanistic art. In addition, Lewis-Williams once met an old woman, probably the last survivor of the south-ern San, whose father had been a shaman. "She demonstrated how dancers seeking power turned to face the paintings on the wall of the rock shelter and how some placed their hands on the paintings of eland to gain power," he says.

The eland, a large antelope, is to the San what horse and bison seem to have been to the Upper Paleolithic people, at least in their art. The eland is the most frequently depicted animal in

San paintings. It has potency, say the San, and it comes in many forms, many qualities. Perhaps the horse and the bison were sources of potency for the Upper Paleolithic people, images that were appealed to and touched when spiritual energy was required.

The question, in effect, is whether Upper Paleolithic art bears the telltale signs of Lewis-Williams's three-stage neuropsychological model, and could thus be shamanistic art. "Upper Paleolithic art includes many of the geometric signs that fall within the range of entoptic elements determined by laboratory research," he says. "Sometimes these motifs are placed on animals, but others, like the grids and fragmented grids at Lascaux, are depicted in isolation. In addition, Upper Paleolithic art includes a range of depictions equivalent to stage three hallucinations: therianthropes, monsters, and realistic animals." The neuropsychological model fits Upper Paleolithic art as well as it does San rock art.

Of the range of images in Upper Paleolithic art, the most arresting are the therianthropes. There are not many of these human-animal figures, but they seize the imagination. The most famous example is the so-called sorcerer in the cave of Trois Frères, in the French Pyrenees. Deep underground, in a cramped cavern, the sorcerer dominates the space. Denis Vialou, who has studied the cave in detail, describes the image: "The body is uncertain, but is some kind of large animal. The hind legs are human, until above the knees. The tail is a kind of canid, a wolf or a fox. The front legs are abnormal, with humanlike hands. The face is a bird's face, odd, with deer's antlers." In a manner unusual for Upper Paleolithic images, the sorcerer is staring directly out of the wall, a full-face stare that transfixes the spectator.

Below the sorcerer are several heavily engraved panels, a riot of animal figures with no apparent order, no pattern. In the midst of all this is another human-animal figure, again with human

The so-called sorcerer from the cave of Trois Frères, in the French Pyrenees, combines human and animal elements. Sometimes thought of as a man dressed up in various animal parts, the image may instead be the kind of human/animal chimera of shamanistic experience.

hind legs. Human hind legs on animals are common in Upper Paleolithic art, incidentally, as are hoofs on otherwise human figures. This therianthrope is standing upright, with a bison's body and the head of a bison, with horns but a somewhat human face. The front legs are odd, in the same way as the sorcerer's forelimbs are. This individual is holding what may be a bow or a musical instrument. "Directly in front of this image is an animal," explains Vialou. "It has reindeer hind legs and rear end, showing

Below the image of the sorcerer in the cave of Trois Frères is a confusion of small engravings (shown at top). In among these French archeologist Denis Vialou has identified a small scene, shown at bottom, that might represent some kind of mythology, perhaps even creation mythology. The human-like figure in the scene is part human, part animal, a powerful image in many mythologies around the world.

female sex prominently displayed, the only one known in Upper Paleolithic art. The rest of the body is bison, the head turned, looking back over its shoulder at the first individual. Something special is going on between these two, I'm sure of that."

We see something similar in Lascaux. The very first beast in the stampede in the Hall of Bulls is an enigma. Known as the Unicorn—wrongly, because it has two very straight horns—this beast has a swollen body on thick limbs and a head of no known animal. There are six circular markings on the body and the partial outline of a horse. Look at the head again, squint—and the profile snaps into that of a bearded man.

These therianthropes in Upper Paleolithic paintings were once dismissed as the products of "a primitive mentality [that] failed to establish definitive boundaries between humans and animals." I do not think so. A more convincing argument is that they represent shamans or hunters dressed in animal skins, sometimes wearing horns or antlers. In the context of shamanistic art, however, they are explained as the outcome of stage three hallucination, something as real for the artist as a horse or a bison.

For the Kalahari San, the eland is the pathway to the potency of the spirit world, a multifaceted symbol of the people's cosmos. When a San shaman goes into a trance, he harnesses that power, becomes part of the world beyond, becomes invisible to the singers and dancers around him, and draws images on the rock face. Ask the San who drew the images, and they say the spirits. The shaman is merely an instrument of the spirits. And the rock face is more than a surface for the paint; it is the boundary of this world and the world beyond. Often in San paintings, a line "disappears" down a crack, to emerge elsewhere, having traversed the spirit world. The rock face therefore becomes part of the meaning of it all, and the rock shelter itself assumes a special status, a place of veneration.

· · ·

There is no doubt in my mind that the caves and rock faces bearing prehistoric images, in Africa and Europe, were special too. Some of them may have been places where bands aggregated, because of a seasonal abundance of a certain food. In that case, the rituals played out there, fragments of which we see in the paintings, built the mythological importance. Some of them may have assumed the status of an aggregation site because a mythological event occurred there. We can be sure that the entire landscape became imbued with elements of mythology, explanations of a people's origin and their place in the world.

Unfortunately, we foreigners are unlikely ever to know the true meaning of the images in the caves and in the rock shelters. Somewhere in Lascaux, I'm convinced, is the entire story of how those Magdalenian people, seventeen thousand years ago, understood their origins. Somewhere—everywhere—in the cave are cryptic messages about how they saw themselves in their world. The place is imbued with meaning, but we can't decipher what is being said. The potency is palpable, but we are culturally blind to its content. In seeking to understand our origins, we come away from a place like Lascaux with a deep conviction of connectedness, and a humility at the power of the human mind.

■

PART SIX

In

Search

of

the

Future

■

19

Origins Reconsidered

T h e   q u e s t i o n I'm asked most frequently when I lecture on human origins is "What will happen next?" The question
itself is at least as significant as any answer that I can offer. The
future, inevitably, is uncertain, and none of us can claim to be a
sure prophet. But the question also flows directly from the very
spirit of humanness that we have been trying to understand.
"What will happen next?"

Three centuries ago, Pascal expressed poignantly this human
questioning about our place in the world: "When I consider the
short duration of my life, swallowed up in the eternity before
and after, the little space which I fill, and even can see, engulfed
in the infinite immensity of spaces of which I am ignorant, and
which know me not, I am frightened, and am astonished at being
here rather than there; for these is no reason why here rather
than there, why now rather than then . . . The eternal silence
of these infinite spaces frightens me."

One does not have to be especially spiritual to experience

awe at the infinity of galaxies we can see in the night sky. Our human consciousness does not merely make possible the question *Why?* It insists that the question be asked. The urge to know is a defining feature of humanity: to know about the past; to understand the present; to glimpse what the future may hold. As Arnold Toynbee said of the impact of subjective consciousness on *Homo sapiens,* "This spiritual endowment of his condemns him to a lifelong struggle to reconcile himself with the universe into which he has been born." The night sky is full of unanswered questions.

In many cases the urge to know surpasses what can be known; questions without answers. Many people, finding this impossible to accept, seek solace in mythic explanations. So attuned is the human mind to look for and find answers that sometimes it extracts meaning where none exists: the face of an old man, a witch, or the image of a monster, seen in an inkspot; psychic portent attributed to mere coincidence; the cry of "why me?" when natural disaster strikes, as if the agent of disaster chooses its victim; the perception of supernatural anger expressed in the violence of an earthquake. Human consciousness demands explanations about the world and is resourceful at creating explanations where none naturally exist.

I labor the point because, while it is a defining element of humanity, the passion to know can easily mislead us. When audiences ask me questions about human fossils, I am aware that they are really asking questions about themselves as members of the human species. This is just as it should be, because paleoanthropology is one of the few sciences that can reach out to touch some of those concerns. But I am also aware that many of the answers I can offer will fall short of expectation, not because they lack content but because the content is disappointingly less noble than may be sought.

The physicist Steven Weinberg said recently that "the more the universe seems comprehensible, the more it also seems pointless." By this he meant that he saw no divine hand guiding the fate of the universe, that the universe proceeds according to its own physical laws, aimed at no ultimate goal. Something similar can be said about *Homo sapiens:* the more we understand about our history, the more our connectedness to nature becomes apparent, the more we see ourselves as part of nature, not apart from nature. Like Weinberg, who still views the universe with awe, I find that an understanding of *Homo sapiens* does not diminish the wonder of our species. While we may be special in many ways, special explanations of our origin and of our place in the universe are not necessary. Our biological heritage is real and tangible, and rooted in the workings of natural selection. "There is a grandeur in this view of life," observed Charles Darwin at the end of *Origin of Species,* referring to the power and creativity of evolution. This is my perspective of human evolution.

As I said in the prologue, I am doubly privileged in the quest to explore the place of *Homo sapiens* in the universe of things. First, my work in searching for and studying human fossils in Kenya has been a personally thrilling and professionally satisfying experience. Few have the opportunity to engage directly in the quest, of leafing back through the pages of history, of seeing traces of our past not previously seen by human eyes. The story to be gleaned from those pages offers a view of humanity that is at times humbling, at times uplifting, always enlightening. It is the perspective of time and change. Second, my involvement in wildlife conservation in Kenya brings an immediacy to our story, a different perspective. I see the extinction of species through human greed and human ignorance. But I also see the wonderful diversity of the natural world. It is a perspective of the power of humanity and, at the same time, of its ultimate insignificance.

There may be some contradictory—even distressing—conclusions and sentiments here. I will explain.

In the fifteen years since I wrote *Origins,* there have been many discoveries, leading to new insights into our past. I feel I now have a clearer view of the place of our species. I want to address three areas of concern here, three issues that I characterize as the Inevitability, the Gap, and the Sixth Extinction. Each of them touches on matters practical and matters intangible. Albert Einstein once quipped that he was interested in finding out "if God had any choice in creating the universe the way He did." In the same vein, I would phrase my quest as finding out what plans, if any, God had for *Homo sapiens.*

First, the Inevitability. I am referring to the common conviction that the arrival of *Homo sapiens* was predestined. The fact that we *are* here contributes, I believe, to this sentiment. It seems to demonstrate that we are here for some purpose; otherwise we would have to concede that we are here by chance, a freak of nature. For many people such a conclusion is unacceptable.

The unacceptability of our being here by chance has been expressed mainly in three ways. First, in the anthropological literature itself, in which the special qualities of *Homo sapiens* are taken to imply that we are here for a purpose and by design. The second is the Anthropic Principle, the view that the universe (and us within it) is the way it is because it could be no other way. And third is a characterization of evolution in which the unfolding of life on earth has followed a path of progress and predictability.

The idea of the Inevitable became incorporated into the anthropological literature in many ways, some blatant, some subtle. For instance, Robert Broom, who in the 1940s and 1950s was responsible for the discovery of many of the hominid fossils in South Africa, was very explicit about it. "Much of evolution looks as if it had been planned to result in man, and in other animals and plants to make the world a suitable place for him to dwell in," he wrote in 1933. "It is hard to believe that the huge-brained thinking ape was an accident." Broom was a convinced Darwinian, but he was so impressed with the "special" qualities of humankind that he felt a spiritual agency must have guided the path of evolution, first preparing the way for *Homo sapiens*, then fashioning the species itself.

Alfred Russel Wallace, co-inventor with Charles Darwin of the theory of natural selection, came to similar conclusions as Broom's. Convinced of the great creative power of natural selection, Wallace nevertheless considered the human mind to be so lofty of intellect and so imbued with moral sense that it was beyond the base world of practical affairs at which evolution operates. He felt much the same about the "soft, naked, sensitive skin of man" and considered that "the structure of the human foot and hand seem unnecessarily perfect for the needs of savage man." He concluded, therefore, that "a superior intelligence has guided the development of man in a definite direction, and for a special purpose."

The arguments of Broom and Wallace are, incidentally, akin to the earlier school of Natural Theology, in which the complexity and beauty of the natural world were taken as evidence of His guiding hand. It was known as the argument from design: the fact that something works well implies that it was designed to be the way it is. William Paley, the greatest exponent of natural theology, drew a famous analogy, of finding a watch on a heath. "The inference, we think, is inevitable; that the watch must have a maker; that there must have existed, at some time, and at some place or other, an artificer or artificers, who formed it for the purpose which we find it actually to answer; who comprehended its construction and designed its use." For every watch there is a watchmaker. So, too, for the perfection of a flower, the sleek, swift horse, and the transcendent human mind.

The constructions of Natural Theology are now part of the history of science, not of current scientific theory, but their attractions can be easily appreciated. So too are the arguments of Wallace and Broom. They were followed by the much revered writings of Pierre Teilhard de Chardin, the French theologian and anthropologist. "Life, if fully understood, is not a freak in the universe—nor man a freak in life," he wrote four decades ago. "On the contrary, life physically culminates in man, just as energy physically culminates in life." This last statement touches closely on the recently developed notion of the Anthropic Principle. "The phenomenon of Man," said Teilhard de Chardin, was "essentially foreordained from the beginning."

The second of the three expressions of the notion of the Inevitability, the Anthropic Principle, is in the domain of physicists, but I believe it resonates with the sentiments described above. Very briefly, cosmologists have become increasingly impressed—and have increasingly expressed their astonishment and awe—that the laws of the universe operate within tight margins with respect to our existence. Change even slightly the fundamental physical forces, and the universe—and life—as we know it could not happen. Everything is precariously balanced, just right, so that we can exist. In order that we shall exist?

Few exponents of the Anthropic Principle go so far as to suggest explicitly, as Teilhard de Chardin did, that man was "essentially foreordained from the beginning." Some, however, come close. For instance, the Princeton theoretical physicist Freeman Dyson says, "I believe that we are here to some purpose, that the purpose has something to do with the future, and that it transcends altogether the limits of our present knowledge and understanding." Others are more cautious. "Without going as far as some," says Martin Rees, a British cosmologist, "I wish to argue that there is something special about the time and place that has produced intelligent life."

The Anthropic Principle can devolve, at its simplest, to the statement that we are here to observe them, therefore the fundamental laws must be as they are. But what of other universes, with other laws? They are unimaginable. There are, without doubt, some interesting philosophical threads running through the fabric of the Anthropic Principle. But I suspect that woven among those threads are strands of "We *are* here, and so in some way it was meant to be."

The sentiments of inevitability expressed explicitly by Teilhard de Chardin, and implicitly and less mystically by others, were replaced by the notion of progress and predictability in evolution, the third medium of expression of the Inevitability. By progress I mean that evolution is viewed as constantly striving for improvement in the biological world, fashioning ever more efficient and successful organisms. The notion of predictability implies that the pattern of life produced by evolution was more or less inevitable, that if the process were set back to the beginning and run again, much the same pattern would result.

The emphasis on progress surely derives from social values, particularly those of Western society, where steady improvement through effort is a virtue to be rewarded. In nature, progress results in the evolution of "higher" forms, a heavily loaded term. Humans, being the "highest" of all forms, are the ultimate product of evolution, or so it goes.

While it is true that, through evolutionary time, forms with superficially greater complexity arose, the fossil record shows no general drive toward improvement, no inexorable progress in the sense that is often meant. Because complexity is built on complexity, the emergence through time of more elaborate forms is a mechanistic inevitability, an evolutionary ratchet. But it is not a general progression. Instead, the record shows a constant stochastic shift in many directions, adaptation to the moment. Through Western eyes, we tend to fixate on the effects of the ratchet and ignore the general pattern.

The notion of predictability, closely allied to progress, is more pertinent to our view of human history. It operates on two levels. The first is manifested in the description of the evolution of a species or group of species. With an adequate fossil record, such as we have for humans, one can build up an account of the changes that have occurred through time. It is then all too easy to describe these evolutionary innovations occurring through a species' history as if they had been steps *on the way toward* that species. Knowing the end of the story, we tell it as if the intermediate steps were written into the script.

The adoption of upright walking, the modification of the dentition, the origin of the expanded brain, the ability to expand the range, the emergence of complex spoken language—each can be seen as part of a cumulative and predictable march toward the present, *Homo sapiens*. One sometimes sees this even in the most technical of anthropological writing. Take, for instance, a recent report in the *American Journal of Physical Anthropology*, concerning the status of Lucy, *Australopithecus afarensis*. "In our opinion *A. afarensis* is very close to a 'missing link,' " wrote the authors of the paper. "It possesses a combination of traits entirely appropriate for an animal that has traveled well down the road toward full-time bipedality." The notion that *afarensis* was traveling anywhere is valid only in hindsight. When it existed, *afarensis* was a successful, stable species, on the way to nowhere.

I too have been guilty of such language, with its assumption of evolution as a scheduled journey. "It was behavioral sophistication that in large measure helped propel this human ancestor along the road to mankind," I wrote in *Origins*, fifteen years ago, referring to *Homo habilis*. "In evolutionary terms the speed at which this journey was undertaken was breathtaking: biological milestones were reached and passed with great rapidity . . . This creature was on its way to modern humans, *Homo sapiens sapiens*." It is a trap, the language of the journey. If we are to reach a true understanding of ourselves and our place in the world, we have to break free of the trap.

I have recounted the history of the human family against a backdrop of climatic and environmental change. Some of these changes, I've suggested, triggered evolutionary innovations among our ancestors. In a thought experiment, we can ask what might have happened if these environmental events had not occurred, or had occurred at different times. What, for instance, might have happened if the drastic global cooling around 2.6 million years ago had not taken place? This cooling, remember, correlates with the origin of new australopithecine species (the robust *boisei*) and the evolution of the enlarged brain, the beginning of *Homo*. With no cooling, and without the ecological modifications resulting from it, perhaps *Homo* would not have appeared then; perhaps not at all.

And what of the environmental and climatic changes associated with the formation of the Great Rift Valley, some ten million years ago and onward? I think it likely that the highland, mosaic environments generated by these events were important in the origin of the hominids in the first place. Had there been no such tectonic events in East Africa at that time, leaving the forests intact, perhaps hominids would not have evolved then; perhaps not at all.

It is incorrect to imagine that a particular species in time has unlimited evolutionary opportunities ahead of it; potential changes are to some degree constrained by its existing anatomical architecture, its historical heritage. *Australopithecus afarensis* was not likely to become evolutionarily transformed into a hoofed grazer, for instance. But it is equally incorrect to assume that the changes that did occur were the only ones that could have happened. Under appropriate conditions of natural selection, *afarensis* could have become a quadrupedal fruit eater, for instance. It didn't; that's all. What happened to this species is a contingent fact of history, not the march down a predestined evolutionary path. It is futile to speculate what might have happened had this or that circumstance been different, but we need to understand that what happened in history was only one of a range of possibilities. *Homo sapiens* was one of a range of possibilities in the evolution of the hominid group, not an inevitable product of that process. As long as we appreciate the true nature of our history, with its uncertainties, we will have a clearer sense of what it is to be a member of the species *Homo sapiens.*

On a larger scale, we have to travel back only some sixty-five million years to find yet another reminder that we are a contingent fact of history, not an inevitability. As most people are aware, sixty-five million years ago marked the end of the Age of Dinosaurs, terminated by some kind of natural catastrophe, almost certainly the collision of Earth with a large asteroid or comet. For 150 million years, dinosaurs were the major terrestrial group, occupying niches that in nature and number were about the same as mammals fill today. The mass extinction that put an end to this group, also devastated other groups, including various forms of mammal. In all, about 60 to 80 percent of all terrestrial species perished in the Cretaceous extinction.

Mammals had existed almost as long as the dinosaurs, but they remained a relatively insignificant part of terrestrial life, occupying a small-bodied, insectivorous niche. Being small, mammals had a better chance of surviving mass extinction (a general rule of the history of life), and many did. Among them was a primitive primate, ancestor to all the six thousand primate species that have existed since that time (about 183 live now). One of the properties of mass extinctions is that many of the normal rules of biology are briefly suspended, principally those relating to everyday competition and survival. Species that survive mass extinctions do so for reasons having to do with geographic distribution, body size, and plain luck. It has nothing to do with inherent superiority or adaptation. Had that primitive primate been less lucky at the Cretaceous extinction, there is no reason to expect that animals like primates would ever have evolved again, no prosimians, no monkeys, no apes—no humans.

This, then, is an important message that comes to us from the fossil record. It makes no difference to the fact that we are here to state that had circumstances been only slightly different in the history of life, we would not be. But it is surely significant to the way we view ourselves to realize that our being here was by no means inevitable, no matter how our very humanness rails against the notion. To answer my earlier question: God surely had no plans for *Homo sapiens*, and could not even have predicted that such a species would ever arise.

The second in my trio of concerns is the Gap: the notion that the very special characteristics the human species enjoys set us apart from the world of nature, "the vastness of the gulf between . . . man and the brutes," as Thomas Henry Huxley put it. Our technological skills, our ability to modify the environment, our cultures, our esthetic sensitivities and ethical sensibilities—all distinguish us from the other species with which we share our world. The gulf seems vast.

Ever since Carolus Linnaeus, in 1758, classified *Homo sapiens* with the rest of the living world, in his *Systema Naturae* (The System of Nature), scholars and theologians alike have tried to insert as much distance as possible between us and the brutes. The reason is obvious: we are special in many ways, and we feel special in a very particular way. From the time the first *Homo sapiens* evolved, and probably somewhat earlier than that, humans have felt in touch not only with the tangible world but also with something beyond, the essences of nature, the spiritual world of their ancestors, the power of the gods. Deriving from the penchant to attribute humanlike motives to nonhuman entities, as I argued in an earlier chapter, the search for meaning in everything, even where none may exist, is the product of our subjective consciousness.

The result has been the elaboration of mythologies to contain and explain the world, religions in many forms. One anthropologist has calculated that, since the beginning of true humanity, more than 100,000 different religions have arisen, most of which have decayed with their creators. "The predisposition to religious belief is the most complex and powerful force in the human mind and in all probability an ineradicable part of human nature," comments the Harvard biologist Edward O. Wilson. "It is one of the universals of social behavior, taking recognizable form in every society from hunter-gatherer bands to socialist republics."

The drive to religion is the urge to explain the unknowable, often by mythic tales and demanding faith. Wilson puts it very nicely: "Men, it appears, would rather believe than know. They would rather have the void of purpose . . . than be void of purpose." Of the many characteristics that we can point to as separating us from the rest of nature, religion is surely among those undeniably unique to the human species.

I am not religious, at least not in the formal sense. As a schoolboy I adopted a kind of personal atheism, and was much ridiculed because my uncle was Archbishop of East Africa at the time. There developed a campaign to "save" me, to which I reacted by being even more adamantly atheistic. I came to be critical of formal religion, particularly of the damage that missionaries were doing to the culture of the people of Kenya. I had no difficulty in accepting the notion that standards of ethics and morality could be derived in the absence of religion. And I now believe that such standards are an inevitable—and predictable—product of human evolution: altruism is part of the behavioral repertoire of social animals, so it can be expected to develop much further in intelligent and intensely social animals, like our human ancestors. This is the humanists' position.

A few years ago, in a town in Minnesota, I gave a lecture on human origins. Afterward an elderly gentleman, a farmer I think, rose and asked a question: "Have you ever met a monkey that knew the meaning of sin, Dr Leakey?" I realized the importance of the question to that gentleman, because the idea of sin is so much part of Western culture. It is a mental concept that helps us guide ourselves and, through ourselves, society in a particular direction. Sin is a human word for knowing wrong from right.

I'm quite sure that monkeys, and apes too, under certain circumstances, know that some things may be unacceptable in social interaction. But monkeys and apes are not burdened with this higher mental concept, sin. I'm sure our more recent ancestors were so burdened, however; it is the product of evolution in the intense social setting of human life.

So, although I am not religious, I can see where the urgent predisposition to religion comes from. An understanding of human history must address this issue, and I think we can do so satisfactorily. The urge to explain, whatever its manifestation—in religion, in philosophy, in science—surely sets a great distance between humans and the other species of today's planet Earth.

So also does culture. *Homo sapiens* is a cultural creature, to an extent and in a manner unmatched by any other species. This extra dimension of behavior essentially creates another world, one that may be constantly reshaped. The generation-to-generation transmission of ideas and knowledge means that we all take part in a cumulative expression of our species. Our view of the world, and the material trappings we enjoy in it, depend in a very direct way on what was done one generation back, ten generations back, a hundred generations back. Today, we are the beneficiaries of our distant ancestors in a way not experienced by any other species.

For perhaps 100,000 years *Homo sapiens* were successful hunters and gatherers, living in small bands, part of larger social and political alliances. Their material worlds were surely limited, but their mythic worlds undoubtedly were rich, and these treasures passed from generation to generation. Then, between twenty thousand and ten thousand years ago, people began to organize their practical lives differently, sometimes exploiting plentiful food resources in a way that allowed less mobility, more stability, perhaps more possessions. Finally, from ten thousand years onward, food production—as against food gathering—became more common, villages sprang up, small towns, cities, city-states, and eventually nation-states. What we call civilization had arrived, founded on generations of slow cultural changes. The range of practical, intellectual, and spiritual possibilities nurtured by civilization is the ultimate expression of the power of culture. Surely it sets us apart from the rest of the species in the world.

I seem to be answering the question of the Gap in positive terms, citing reasons why it exists. In one very real sense, Huxley was right in saying that "a vast gulf" exists between us and the brutes. The products of subjective consciousness and the products of culture seem to attest to that. Indeed, so great a gap did Julian Huxley, Thomas Henry's grandson, perceive between humans and the rest of the animate world that he suggested *Homo sapiens* be classified within an entirely new grade, the Psychozoan. "The new grade is of very large extent, at least equivalent in magnitude to all the rest of the animal kingdom," he suggested in 1958, "though I prefer to regard it as covering an entirely new Sector of the evolutionary process, the psycho-social, as against the entire nonhuman biological Sector."

To make humans the only member of a third, distinct kingdom in the world of nature—Animals, Plants, and Psychozoans —is about as far as one could go in placing a gap between us and the rest of nature. But one of the most pertinent lessons we learn about ourselves by learning about our past is that this gap is something of an illusion, an accident of history. We feel special and separate because no species comes close to our accomplishments. And yet, if we look to the fossil record, we see the links in the chain that binds us to the rest of nature.

The links are not just nominal; they are the species of hominid through which we trace our lineage, finally, back to a common ancestor with the apes, an unbroken genetic link to the nonhuman world of nature. I have argued that the qualities we identify as defining humanity—consciousness, compassion, morality, language—emerged gradually during our history. They did not arise suddenly and late. Had the human brain been suffused with consciousness, compassion, morality, language only with the origin of *Homo sapiens*, and our earlier forebears been little more than erect apes, then there would be some merit in the claim that humans are indeed separate from all of nature. But because the emergence was gradual, other species do indeed match up to us to some degree. These species just happen no longer to be here; again, a contingent fact of history. The Gap is not as vast as Huxley—grandfather and grandson—believed. Indeed, it is closed.

Now to the third of my three concerns: the Sixth Extinction. Here, I want to address our short-term behavior and our long-term prospects. One is uncertain, the other not.

In my day-to-day work as director of Kenya Wildlife Service I face the practical issues of trying to prevent the extinction of species. Sometimes this demands deploying a heavily armed antipoaching squad to protect elephant and rhinoceros from slaughter; sometimes it requires saving land from encroaching farms so that a rare bird can survive. Always there is the press of human desires and needs against a shrinking natural world, the growing human population against dwindling populations of wild creatures.

In much of the world, the same process is occurring; natural lands are sacrificed to expanding human populations and the rapacious appetite of economic development. The process is already complete in much of Europe and North America, of course, so the focus of concern is thrown on less developed regions. The world's existing biodiversity is being steadily eroded; species are being pushed into extinction at an increasingly rapid rate. According to some estimates, as much as 50 percent of the world's species will be extinct within three decades. That is a figure of some considerable significance, not just in numbers but in the overall history of life on Earth.

Since the origin of complex forms of life on Earth, there have been five mass extinctions, events during which the number of living species collapsed catastrophically. They are Ordovician, 430 million years ago, the Devonian, 350 million years ago, the Permian, 225 million years ago, the Triassic, 200 million years ago, and the Cretaceous, 65 million years ago. (There are a number of smaller events scattered among and after these, which gives a rough periodicity of a significant extinction every twenty-six million years.) Each time one of these mass extinctions (known to paleontologists as the Big Five) struck, it essentially changed the Earth's biota, and on a grand scale. The Permian event, for instance, wiped out 96 percent of all species, close to a complete rout.

Periodic mass extinctions therefore characterize Earth history, as do the rapid recoveries from these events. After each collapse, ecological opportunities were present for the survivors, and history shows that they exploited these opportunities thoroughly and swiftly. Over a few tens of millions of years the overall diversity often reached close to or even exceeded the levels before the mass extinctions. And, until recent times, following the Cretaceous extinction, the level of diversity was higher than at any time in Earth history. One therefore sees a sawtooth pattern of Earth history: a sharp downward slope (extinction) followed by slow upward slope (recovery), repeated time and time again.

Although subjectively it is difficult to discern, we are now in the midst of the Sixth Extinction. The loss of 50 percent of species is rightful definition of the title. No catastrophic impacts from asteroids this time, no massive chains of volcanic eruption, no global disaster of natural origin—just the inexorable growth of human populations, enveloping and destroying the habitat of the rest of the world's organisms. A mass extinction of unique origin but with familiar effects: the number of extant species is collapsing, and we are the agents of its demise.

I've heard some paleontologists, comforted by the fossil record, argue that we should not concern ourselves; mass extinctions have occurred before, and the biota has always recovered.
It will recover again, they say. Others disagree. The cause of
extinction in previous events was always transitory, so recovery
was possible. Not this time. This time the destructive agent is
here to stay; recovery will not be possible. Deciding which of
these two viewpoints is correct forces us to put *Homo sapiens* into
the largest perspective of all: how long it will survive. "What will
happen next?"

One of the lessons we draw from the fossil record is that, for
the most part, species do not last very long. On average, inverte-
brate species have a longevity of five to ten million years, for
instance, and the figure for vertebrates is about two million
years. As a result, more than 99 percent of all species that have
ever lived are now extinct. Species go extinct for the most part
not because they are in some way inferior, but because they
succumb to the vagaries of mass extinction. What of *Homo sapiens?*

Our species is relatively young, not much more than 100,000
years old. And, if the periodicity of the extinction pattern holds
up, the next major event is not due for another twelve million
years or so. The prospects, then, look good. Or do they? Even if
our species is able to avoid self-destruction, either spectacular
through military conflagration or slow through environmental
strangulation, and were to live out the average vertebrate's span,
we still face the prospect that there will one day be an Earth with
no *Homo sapiens* on it.

It has not been the fate of all species to become extinct
without issue, of course. Some are transformed by evolution into
descendant species. Could this be the prospect for *Homo sapiens*,
ancestor of *Homo technologicus*, for example? How can we know,
history being as capricious as it is? But my guess is no.

Culture, which so transformed and enriched the life of *Homo sapiens*, may also block its further evolution. Evolutionary change by natural selection proceeds by the differential survival of genetically favored individuals. Culture effectively eliminates that process, by making survival subject to many nongenetic factors. Unless there is genetic intervention by new technology or deliberate breeding programs over many thousands of years—both of which options properly ring ethical alarm bells in our society— further evolution of *Homo sapiens* is probably at an end.

We should, however, put any notion of further evolution of *Homo sapiens* in a larger time perspective. In our recent history, two intellectual revolutions shook humanity's perception of itself. The first was the Copernican revolution, beginning in the sixteenth century, which dislodged the Earth, and therefore humans, from the center of the visible universe to the position of a small planet, with others, circling a small sun. The second was the Darwinian revolution, which placed humans in the same biological rule book as the Earth's other species. To these two insults to humanity's perception of itself in the universe of things there has been added, only recently, a third: the scale of the universe itself.

Some twenty billion light years across, the universe is now seen to be unimaginably vast, our solar system and its host galaxy, the Milky Way, an insignificant corner of infinite time and space. Such a perspective surely challenges the strength of the human spirit in its perception of itself. Our own sun, it is calculated, will last another five or ten billion years, producing heat and light sufficient to sustain life on Earth. Figures like these cheat the mind in its attempt to comprehend them. But it is a certain guess that, long before the sun's energy is finally spent, *Homo sapiens* will no longer exist, another extinct species in Earth's history of biotic collapse and recovery. The Earth will go on in the absence of *Homo sapiens*, evolution and extinction playing against each other for another few billion years. Perhaps the

mass extinction some seven hundred million years after the

beginning of complex life on Earth, the Sixth Extinction, will come to be seen (by whom?) as an aberrant blip in Earth history, a temporary reduction in species richness, out of sequence in time.

I agree, therefore, with the paleontologists who argue that, in the end, the Earth's biota will recover from its human-inflicted collapse, the Sixth Extinction. The continents will continue to drift around the globe as they always have, driven into collision and torn apart again by massive tectonic forces. The organisms of the land, the sea, and the air will go through alternating phases of mass extinction and recovery, evolutionary innovation producing new classes that we can hardly begin to guess at: amphibians, reptiles, mammals . . . what next? Continuous variation on established ecological themes; that's how evolution works through time.

It is possible, perhaps even likely, that over the eons of evolutionary time, intelligent life forms will emerge again, consciousness reborn on Earth. Such an organism would no doubt try to make sense of the earlier civilization that existed on the planet, try to piece together from meager archeological remnants the way that civilization lived and what brought it to an end.

Fantasy, of course, but a useful one in forging a true perspective of *Homo sapiens*. It has been said that we humans, with our intelligence, our technology, and our power, are the stewards of planet Earth—that its future is in our hands. Because we find it impossible, individually and collectively, to imagine a time when we will no longer exist, we naturally equate the future of *Homo sapiens* with the future of the planet. But the logic of the fossil record, and the logic of a true understanding of *Homo sapiens* as one species among many, forces us to accept that this is not the case. We are not stewards of the Earth, forever and a day. We are merely short-term tenants, and pretty unruly and destructive ones at that.

357

Now, although I agree with the paleontologists who argue that the biota will recover from the Sixth Extinction, I do not concur that it is therefore of no real concern what we do while we are here. An understanding of human origins tells us that *Homo sapiens* is a part of the natural world here on Earth, one species among many. But we do have the intelligence to comprehend the impact of what we do as a species on the rest of the species around us.

The ecosystem of which *Homo sapiens* is a part is a complex entity, one that is at once robust and fragile. Perturb it by natural catastrophe—hurricane, fire, or volcanic eruption, for instance—and the immediate result is apparent devastation. But before long it bounces back. Think of the green shoots among the ashes of the Yellowstone Park fire, the spring after that massive conflagration. Think of the emerging life around the slopes of Mount St. Helens, not many years after its catastrophic explosion. And the fossil record shows the biota's same resilience after destruction on a global scale, the mass extinctions. The scale against which we should judge the impact so far of *Homo sapiens* on other species—the scale against which we should contemplate our responsibilities as tenants of Earth—is more like that of the Mount St. Helens eruption than mass extinction events and the subsequent recoveries. Rapidly, however, we are fueling the engines of the Sixth Extinction.

What, then, should be our concern? I believe that the qualities of humanness—consciousness, compassion, morality, language—arose gradually in our history, products of the evolutionary process that shaped our species. These qualities are, of course, most appropriate in the interactions among individual humans; they are the threads that hold social fabric together. But, together with our creative intellect, they form part of our perception of the rest of the world of nature. I am not suggesting, as some people do, that every species of plant or animal has the same rights in society as humans. It is correct that we recognize the special value of a human life. But it is also correct that we recognize the place in nature of the human species, *Homo sapiens*, one species among many. This is the true insight into our origins.

It matters not at all that other species do not possess a degree of consciousness like ours, do not experience feelings in the way we do. They are part of our world; we are part of theirs. Our greater intellect may confer on us an enhanced ability to exploit the natural resources of the world. But—and I feel this very strongly—it also lays on us an enhanced responsibility to husband those resources carefully, to be sensitive to the knowledge that a species, once extinct, is destroyed forever. By impoverishing the environment, we impoverish our own lives, in this short-term tenancy we have on planet Earth.

"What will happen next?" The world will carry on without us once *Homo sapiens* becomes extinct, no doubt about that. But that is of no account to me, to my children, and to the rest of the human species. None of us will be around. The period over which we have responsibility, the period in which we have an interest as a species, the period in which we can make a difference, is now. We need to be quite clear about ourselves as one species among many. We need to be better tenants. I hope that what will happen next is that, collectively, we decide we *will* be better tenants.

Origins Reconsidered, the title of this chapter and of the book, has been a personal odyssey, a journey of discovery that may be close to an end for me. The privilege and responsibility of being involved in the search for clues to our past, the privilege and responsibility of fighting battles for wildlife conservation, both have combined to create a personal experience as profound as I could have hoped for. I expect to continue my involvement in wildlife conservation for some years, but I may never again be as involved as I once was in the search for human ancestors. The journey of discovery has taken me to new territories, territories from where the place of *Homo sapiens* in the universe of things is more clearly perceived.

I have learned that our future is here, now. We are already living it.

Index

361

About the Authors

RICHARD LEAKEY is the world's most famous living paleoanthropologist. He resigned from his position as chairman of the National Museums of Kenya when Kenya's President, Daniel arap Moi, appointed him to head the Kenya Wildlife Service.

ROGER LEWIN coauthored *Origins* with Richard Leakey and is the successful author of several prize-winning science books.